Springer Series in Adaptive Environments

Editors-in-Chief

Holger Schnädelbach⊙, Mixed Reality Laboratory, University of Nottingham, Nottingham, United Kingdom

Henriette Bier⊙, Robotic Building, Delft University of Technology, Delft, The Netherlands; Anhalt University of Applied Sciences, Dessau, Germany

Kristof van Laerhoven⊙, Ubiquitous Computing, University of Siegen, Siegen, Germany

The Springer Series in Adaptive Environments presents cutting-edge research around spatial constructs and systems that are specifically designed to be adaptive to their surroundings and to their inhabitants. The creation and understanding of such adaptive Environments spans the expertise of multiple disciplines, from architecture to design, from materials to urban research, from wearable technologies to robotics, from data mining to machine learning and from sociology to psychology. The focus is on the interaction between human and non-human agents, with people being both the drivers and the recipients of adaptivity embedded into environments. There is emphasis on design, from the inception to the development and to the operation of adaptive environments, while taking into account that digital technologies underpin the experimental and everyday implementations in this area.

Books in the series will be authored or edited volumes addressing a wide variety of topics related to Adaptive Environments (AEs) including:

- Interaction and inhabitation of adaptive environments
- Design to production and operation of adaptive environments
- Wearable and pervasive sensing
- Data acquisition, data mining, machine learning
- Human-robot collaborative interaction
- User interfaces for adaptive and self-learning environments
- Materials and adaptivity
- Methods for studying adaptive environments
- The history of adaptivity
- Biological and emergent buildings and cities

More information about this series at http://www.springer.com/series/15693

Henriette Bier

Editor

Robotic Building

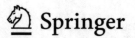 Springer

Editor
Henriette Bier ⓘ
Delft University of Technology
Delft, The Netherlands

and

Anhalt University of Applied Sciences
Dessau, Germany

ISSN 2522-5529   ISSN 2522-5537   (electronic)
Springer Series in Adaptive Environments
ISBN 978-3-030-10000-1   ISBN 978-3-319-70866-9   (eBook)
https://doi.org/10.1007/978-3-319-70866-9

Advisory/Reviewer Committee

Keith Green (Cornell University)
Justin Dirrenberger (CNAM)
Sebastian Vehlken (Leuphana University)
Jean Vanderdonkt (Université catholique de Louvain)
Mikael Wiberg (Umeå University)
Omar Khan (Buffalo University)
Marcus Foth (Queensland University of Technology)
Ava Fatah (University College London)
Martin Knöll (Technical University Darmstadt)
Gerd Kortuem (Technical University Delft)
Hedda Schmidtke (University of Oregon)
Norbert Streitz (Smart Future Initiative)
David Gerber (Arup Research)
Philippe Morel (ENSAPM)

Preface

While architecture and architectural production is increasingly incorporating aspects of non-human agency employing data, information and knowledge contained within the (worldwide) network connecting electronic devices, the relevant question for the future is not if but how robotic systems will be incorporated into building processes and physically built environments (Bier 2013) in order to improve everyday life. The first book of the Adaptive Environments (AE) series on Robotic Building (RB) aims to answer this question by critically reflecting on the achievements of the last decades in the application of robotics in architecture and furthermore outlining potential future developments and their societal implications. The focus is on robotic systems embedded in buildings and building processes implying that architecture is enabled to interact with its users and surroundings in real time and corresponding Design-to-Production and -Operation (D2P&O) chains are (in part or as whole) robotically driven.

Such modes of production and operation involve agency of both humans and non-humans. Thus, agency is not located in one or another but in the heterogeneous associations between them (Bier 2016), and authorship is neither human nor non-human but collective, hybrid and distributed (Latour 2014).

Robotic Building (RB) as investigated in this AE volume relies on interactions between human and non-human agents not only at design and production level but also at building operation level, wherein users and environmental conditions contribute to the emergence of multiple architectural configurations. In this context, design becomes process—instead of object-oriented—use of space becomes time—instead of program- or function-based—which implies that architects' design increasingly processes, while users operate multiple time-based architectural configurations emerging from the same physical space that may physically or sensorially reconfigure in accordance with the environmental- and user-specific needs. If spatial reconfiguration may be facilitating multiple, changing uses of physically built space within reduced timeframes, interactive energy and climate control systems embedded in building components and employing renewable energy sources, such as solar and wind power, may reduce architecture's ecological footprint while enabling a time-based, demand-driven use of space. Both rely on virtual modelling and simulation interfacing the production and real-time operation of physically built space (Latour 2014) establishing thereby an unprecedented Design-to-Robotic-Production and -Operation (D2RP&O) feedback loop, which is a focus of this book.

The integration of D2RP with D2RO implies understanding both approaches as requiring safe human–robot interaction and collaboration in the production and operation of buildings. Since both production and operation of buildings take place in more or less unstructured environments, both imply similar challenges and opportunities. RB links, therefore, design and production with smart operation of the built environment and advances applications in performance optimization, robotic manufacturing and user-driven building operation.

Delft, The Netherlands/Dessau, Germany   Henriette Bier

References

Bier H (2013), Robotic(s in) architecture. Interactive Architecture #5. Jap Sam Books, Heijningen, p 6–8
Bier H (2016) Robotic Building as integration of Design-to-Robotic-Production and -Operation. Next Generation Building #3. TUD, Delft, p 1–5
Latour B (2014) Reassembling the social: an introduction to actor-network-theory. Oxford University Press, Oxford, p 63–86

Acknowledgements

The first volume of Springer's Adaptive Environments book series has profited from the Robotic Building session at the Game Set Match #3 symposium organized at TU Delft (2016). It builds up on the abstracts presented at this session, which were published in Spool's first issue on Cyber-physical Architecture (2017). It has profited from the contribution of reviewers, who have single-blind reviewed each chapter. In particular, the contribution of the editors-in-chief Holger Schnädelbach and Kristof Van Laerhoven and of the editorial board Keith Green (Cornell University, USA), Justin Dirrenberger (CNAM, France), Sebastian Vehlken (Leuphana University, Germany), Jean Vanderdonkt (Université catholique de Louvain, Belgium), Mikael Wiberg (Umeå University, Sweden), Omar Khan (Buffalo University, USA), Marcus Foth (Queensland University of Technology, Australia), Ava Fatah (University College London, UK), Martin Knöll (Technical University Darmstadt, Germany), Gerd Kortuem (Technical University Delft, Netherlands), Hedda Schmidtke (University of Oregon, USA), Norbert Streitz (Smart Future Initiative, Germany), David Gerber (Arup Research, UK) and (Philippe Morel, ENSAPM, France) needs to be acknowledged. Furthermore, the contribution of Senatore Gennaro (EPFL, Switzerland), Kevin Clement (Kengo Kuma and Associates, Japan), Tapio Heikkilä (VTT, Finland), Christian Karl (University Duisburg, Germany) and from Springer's side Beverley Ford (Springer Computer Science, UK) and Nancy Wade-Jones (Springer Nature, UK) requires additional acknowledgement.

Contents

Editor and Contributors

About the Editor

Henriette Bier graduated (1989) from the University of Karlsruhe in Germany. She worked with Morphosis (1999–2001) on internationally relevant projects in the USA and Europe. She has taught computational design (2002–2003) at universities in Italy, Austria, Germany, Belgium and the Netherlands. Since 2004, she has been teaching and researching as a Ph.D. researcher and later on as Assistant and Associate Professor at the Technical University Delft (TUD) with a focus on application of robotics in architecture. In 2017, she was appointed Professor at Dessau Institute of Architecture in Germany.

From 2005–06, she initiated and coordinated the workshop and lecture series on Digital Design and Fabrication with invited guests from MIT and ETHZ. Since 2006, she co-developed the education and research frameworks for Non-standard and Interactive Architecture at Hyperbody and Border Conditions at Public Building, TU Delft. Since 2014, she has been leading the Architecture M.Sc. specialization with a focus on robotics in architecture, covering 9 courses of 80 ECTS offered to about 70 students per year. She co-tutored more than 100 graduation projects from which several have received Archiprix nominations, mentions and prizes with the most prestigious one being the 1st International Archiprix prize in 2015. Her graduates successfully practice architecture in internationally known offices (such as Foster, Hadid, UN Studio, OMA) and implement research at relevant institutions (such as TUD, ETHZ, US).

In 2008, she finalized her Ph.D. on System-embedded Intelligence in Architecture, and coordinated until 2010 two EU-funded projects focusing on F2F and online postgraduate education. In 2010, she started developing the academic education and research framework for Robotic Building (RB). In 2011, she joined Delft Robotics Institute (DRI), and in 2013 she received 4TU funding, which allowed her to set up the first robotic laboratory in the faculty of Architecture and the Built Environment, TU Delft, with a team of six researchers. In 2016, she received again 4TU funding and together with funding from the industry she secured 2017 an additional position for a Ph.D. student working on multi-material robotic 3D printing.

She is a member of the M.Sc. admissions committee since 2013 and was research coordinator of the Architectural Engineering and Technology (AET) department from 2014–16. In this role, she has been leading the development of the online Wiki platform to support the AE&T research. She has been a member of several Ph.D. admission committees and is now chairing the Ph.D. admissions committee of AET.

Results of her research are published internationally in more than 120 journals, books and exhibits. Her media appearances such as the TEDx held in Delft 2015 and her participation in exhibits such as Imprimer le Monde at Centre Pompidou in Paris 2017 ensure dissemination of her research to larger audiences. She is a member of several editorial boards and scientific committees and co-organized conferences and co-edited several books and journal issues on data-driven design and robotics in architecture. Most recently, she co-founded the International Association Adaptive Environments with its publication platform in the Springer book series Adaptive Environments.

Contributors

Jari M. Ahola VTT Technical Research Centre of Finland, Oulu, Finland

Ana Anton Faculty of Architecture and the Built Environment, TU Delft, Delft, The Netherlands

Hendro Arieyanto Casajardin Residence Software Maintenance, Jakarta Raya, Indonesia

Henriette Bier Faculty of Architecture and the Built Environment, TU Delft, Delft, The Netherlands; Dessau Institute of Architecture, HS Anhalt, Dessau, Germany

Serban Bodea Faculty of Architecture and the Built Environment, TU Delft, Delft, The Netherlands

Hadin Charbel Advanced Design Studies, University of Tokyo, Tokyo, Japan

Kevin Clement Advanced Design Studies, University of Tokyo, Tokyo, Japan

Justin Dirrenberger Laboratoire PIMM, Arts et Métiers-ParisTech, Cnam, CNRS, Paris, France; XtreeE, Rungis Cedex, France

Keith Evan Green Cornell University, Ithaca, NY, USA

Tapio Heikkilä VTT Technical Research Centre of Finland, Oulu, Finland

Pekka Kilpeläinen VTT Technical Research Centre of Finland, Espoo, Finland

Jiang Lai Advanced Design Studies, University of Tokyo, Tokyo, Japan

Alexander Liu Cheng Faculty of Architecture and the Built Environment, TU Delft, Delft, The Netherlands

Deborah Lopez Advanced Design Studies, University of Tokyo, Tokyo, Japan

Timo Malm VTT Technical Research Centre of Finland, Tampere, Finland

Sina Mostafavi Faculty of Architecture and the Built Environment, TU Delft, Delft, The Netherlands; Dessau Institute of Architecture, HS Anhalt, Dessau, Germany

Yusuke Obuchi Advanced Design Studies, University of Tokyo, Tokyo, Japan

Timo Salmi VTT Technical Research Centre of Finland, Espoo, Finland

Jun Sato Advanced Design Studies, University of Tokyo, Tokyo, Japan

Holger Schnädelbach Mixed Reality Lab, School of Computer Science, University of Nottingham, Nottingham, UK

Gennaro Senatore Applied Computing and Mechanics Laboratory (IMAC), School of Architecture, Civil and Environmental Engineering (ENAC), Swiss Federal Institute of Technology (EPFL), Lausanne, Switzerland

Doris K. Sung School of Architecture, University of Southern California, Los Angeles, USA; DOSU Studio Architecture, Los Angeles, USA

Sebastian Vehlken Institute for Advanced Study on Media Cultures of Computer Simulation, Leuphana University Lüneburg, Lüneburg, Germany

Abstract

The first volume of the Adaptive Environments Springer book series focuses on *Robotic Building*, which refers to both physically built robotic environments and robotically supported building processes. Physically built robotic environments consist of reconfigurable, adaptive systems incorporating sensor–actuator mechanisms that enable buildings to interact with their users and surroundings in real time. These require Design-to-Production and -Operation chains that are numerically controlled and (partially or completely) robotically driven. From architectured materials, on- and off-site robotic production to Robotic Building operation augmenting everyday life, the volume examines achievements of the last decades and outlines potential future developments in Robotic Building.

Keywords Architecture · Adaptation · Reconfiguration · Robotic Building Design-to-Robotic-Production · Design-to-Robotic-Operation

Introduction

The first book of Springer's Adaptive Environments (AE) book series aims to answer the question of how robotic systems are incorporated into building processes and physically built environments (Bier 2013) in order to improve building production and operation processes by critically reflecting on the achievements of the last decades and furthermore outline potential future developments and their societal implications. The focus is on robotic systems embedded in buildings and building processes implying that architecture is enabled to interact with its users and surroundings in real time, and corresponding Design-to-Production and -Operation (D2P&O) chains are (in part or as whole) robotically driven (Bier 2017).

Robotic Building (RB) as investigated in this first AE volume relies on interactions between human and non-human agents at design, production and building operation level. The latter implies that users and environmental conditions contribute to the emergence of various architectural configurations. These changing configurations allow spaces to adapt to variations in occupancy and use, climate needs, etc. They rely on virtual modelling and simulation interfacing the production and real-time operation of physically built space establishing thereby an unprecedented Design-to-Robotic-Production and -Operation (D2RP&O) feedback loop, which is a focus of this book (Bier 2017; Bier and Knight 2014).

The chapters of the first volume of the Adaptive Environments Springer book series address the D2RP and D2RO aspects from various perspectives. For instance, *robot–robot interaction* and *human–robot collaboration* are investigated with respect to their potential to improve productivity, while *robotics embedded in built structures* is explored from the perspective of adaptation to structural, environmental or functional requirements. Materials are discussed from the perspective of *smart and architectured materials* as approaches to work with even design material properties in order to efficiently produce and operate buildings.

Sebastian Vehlken discusses and reviews in Chap. 1 the application of swarm intelligence (SI) and swarm robotics (SR) to architecture and building construction from a history of science and technology perspective. He explores the conceptual entanglements of swarm intelligence and provides a critical overview of seminal SI approaches in architectural design. SR is investigated mainly from the perspective of *robot–robot interaction* and its potential for construction processes. From an applied science perspective, Timo Salmi et al. present in Chap. 2 *human–robot collaboration*. Safety and control technologies of human–robot collaboration are outlined, and sensor-assisted control approaches for industrial robots are described in detail. Furthermore, applicability of sensor-based robotics and potential of robotics in building construction in general are also evaluated. Chap. 3 focuses on reexamining and exploring through history, thus precedents, and case studies the relevance and potential of *intertwining human and non-human agents* in architectural production.

Chapter 4 by Justin Dirrenberger explores *architectured materials* as bridging across the micro-scale of materials to the macro-scale of architectural structures, with the ultimate goal to implement large-scale *robotic additive manufacturing* at building scale. Robotic additive manufacturing is addressed in the next chapter as well but more from the perspective of its potential to complement other *robotic building* techniques required in the production and operation of buildings. Henriette Bier et al. explore challenges and opportunities of integrating the two into a chained D2RP&O process.

D2RO is furthermore explored in Chap. 6 authored by Keith Green, who argues that by *embedding robotics in buildings* interactive and therefore intimate relationships between physical environments and humans are forged. These rely, according to Holger Schnädelbach, on feedback loops between humans and environment that shape such interactions. He examines in the next chapter human–machine interaction and requirements for *adaptive architecture*. In terms of their efficiency, Senatore Gennaro identifies a design approach for *adaptive structures* that is using a strategically integrated actuation system, which redirects the internal load path to homogenize the stresses and to keep deflections within limits by changing the shape of the structure. In contrast, Doris Sung identifies *passive–active systems* as reasonable alternative and complement to the growing number of active systems that are imbued with artificial intelligence.

The chapters of the first volume of the Adaptive Environments Springer book series present theoretical and applied research on robotics in architecture and building construction, identifying its challenges and opportunities. Main consideration is that production and operation of buildings are in the future robotized. Thus understanding that certain skills sets are better acquired and executed by humans while others by machines is understood as key to developing future interaction scenarios between humans and robots. The goal is to take advantage of robotics when it comes to heavy work, precision, repeatability, etc., while still relying on human common sense, creativity, decision-making, etc. The expectation is that in future, interaction between robots and humans at building production and operation level will increase and diversify, with robots becoming more autonomous and human–robot teams collaboratively sharing control. These interactions are *adaptive* in that the individual and collective behaviours reconfigure according to the changing environment.

References

Bier H (2013), Robotic(s in) architecture, Interactive Architecture #5. Jap Sam Books, Heijningen, p 6–8

Bier H (2017) Robotic Building as integration of Design-to-Robotic-Production and -Operation, Cyber-physical Architecture #1. TUD, Delft

Bier H, Knight T (2014) Dynamics of data-driven design, Footprint #15. TUD, Delft, p 1–4

Chapter 1
Visions of Process—Swarm Intelligence and Swarm Robotics in Architectural Design and Construction

Sebastian Vehlken

Abstract This chapter discusses and reviews the application of swarm intelligence (SI) and swarm robotics (SR) to architecture and construction from a history of science and technology perspective. In a first step, it explores the conceptual entanglements of swarm intelligence and adaptive environments and situates them in the context of a recent theoretical discourse about "media ecologies". The second part provides a critical overview of seminal SI approaches for architectural design. These scrutinize novel connections between architecture as a site of material composition and as a site of spatial practices by computer experiments in software environments. Its guiding hypothesis is that SI technologies here are primarily used to create *diversity*. Subsequently, the third part of the chapter examines in which ways recent advances in collective robotics lead to further materializations of the adaptive capabilities of swarming that go beyond software applications. It presents three state-of-the-art examples of SR for architectural construction and demonstrates that SR in architectural construction—in contrast to the paradigm of *diversity* discussed in the context of architectural design—work best in context with a high degree of standardization and pre-defined modularization, or, on the basis of *regularity*.

1.1 Introduction

Swarm Intelligence (SI) has inspired—and sometimes haunted—architectural thought and architectural design for more than two decades. In 1994 Kevin Kelly, at that time editor of Wired Magazine, enthusiastically embraced Mark Weiser's (1991) vision of ubiquitous computing devices:

S. Vehlken (✉)
Institute for Advanced Study on Media Cultures of Computer Simulation,
Leuphana University Lüneburg, Lüneburg, Germany
e-mail: sebastian.vehlken@leuphana.de

© Springer International Publishing AG, part of Springer Nature 2018
H. Bier (ed.), *Robotic Building*, Springer Series in Adaptive Environments,
https://doi.org/10.1007/978-3-319-70866-9_1

[A]s chips, motors, and sensors collapse into the invisible realms, their flex-ibility lingers as a distributed environment. The materials evaporate, leaving only their collective behavior. We interact with the collective behavior—the superorganism, the ecology—so that the room as a whole becomes an adaptive cocoon. (Kelly 1994: 150).

As of today, we realize that such 'superorganisms'—at least at the consumer end—are called Alexa or Siri, and that behind the distributed devices of such ambient and adaptive intelligences lurk the monpolistic and centralist data mining forces of tech giants: the data leeches behind the swarm. Ten years after Kelly and Weiser Kas Oosterhuis (2006) more specifically described the potentials of swarming for a renovation of traditional architectural approaches in a dawning age of digital networks and tools. Surrounded by the emerging accessibility of open source and free software his *Swarm Architecture* manifesto on the one hand became a conceptual framework that conceived of buildings as dynamic point clouds which mesh a multitude of building elements, inhabitants, and their actions (see also Friedrich 2009), whilst on the other called for novel collaborative work modes facilitated by digital technologies. It spawned a number of experimental architectural buildings which involved SI software applications, e.g. *ONL's* 'Water Pavilion', or *Laboratory for Visionary Architecture's* 2014 pavilion for Philips Lighting (LAVA 2014), and has been extended by Studio *Kokkugia* (2010) from buildings to cityscapes—architecture theorist Neil Leach called this *swarm urbanism* (Leach 2009). However, only recently such conceptual and computational SI approaches to architecture began to leave their software environments and spawned real-life cousins (see e.g. Wiesenhuetter et al. 2016): Research projects like the termite-inspired *TERMES* at Harvard University (see Petersen 2016; Petersen et al. 2011; Werfel et al. 2006; Werfel et al. 2014) or the *Aerial Robotic Construction* group of ETH Zurich which makes use of flocking algorithms (see Augugliaro et al. 2013; Willmann et al. 2012) started engineering robot collectives for actual architectural construction.

No matter whether ideas of using SI in architecture rose from wet dreams of tech advocates or concern concrete engineering problems, they refer to a particular mindset of creating viable solutions for multi-dimensional or opaque problem spaces by benefiting from the capacities for self-organization of collectives of rather simple, but highly relational individual agents. SI is grounded in the idea that the complex adaptive behavior of a system at the global level can be effected by multiple parallel interactions of very simply constructed individuals at the local level which follow a set of only a few behavioral rules. Figure 1.1 Compelling cases are the three steering rules of avoidance (avoid collision with local flock mates), alignment (steer towards the average heading of local flock mates), and cohesion (steer towards the locally perceived center of the flock) which one finds in bird flocks or fish schools, or communication through stigmergic signs which individuals leave in the environments like in some types of social insects. Such collectives possess certain abilities that are lacking in their component parts. Whereas an individual member of a swarm commands only a limited understanding of its environment, the collective as a whole is able to adapt nearly flawlessly to the changing conditions of its surroundings. Without recourse to an overriding authority or hierarchy, such collectives organize themselves quickly, adaptively, and uniquely with the help of their distributed control

Fig. 1.1 In 1986, computer graphics designer Craig Reynolds developed a pioneering SI application known as the Boids Simulation. Its 'bird-oid' agents show self-organized collective movement based on a flocking algorithm of only three basic behaviors in local neighborhoods: Separation (steer to avoid crowding local flock mates), Alignment (steer towards the average heading of local flock mates), and Cohesion (steer to move toward the average position of local flock mates). The screenshots are taken from the graphic console of a Symbolics Lisp Computer. (Reynolds 1987)

logic. Within swarms, the quantity of local data transmission is converted into new collective qualities.

The epistemological foundations of that particular mindset, however, are more intricate than the usual bionic narrative of bio-inspired technical systems. Swarms, flocks and schools first emerged as operational collective structures by means of the reciprocal computerization of biology and biologization of computer science. In a recursive loop, swarming in social insects, flocking birds or schooling fish inspired agent-based modelling and simulation (ABM), which in turn provided biology researchers with enduring knowledge about their dynamic collectives. This conglomerate led to the development of advanced, software-based 'particle systems'.

Agent-based applications are used to model solution strategies in a number of areas where opaque and complex problems present themselves. Swarm intelligence (SI) has thus become a fundamental cultural technique for governing dynamic processes (see Vehlken 2013).

Distributed, leaderless, robust, flexible and redundant, swarms adapt swiftly to changing environmental forces. Moreover, they form a specific secondary environment, which surrounds the swarm-individuals and facilitates adaptive processes by way of rapid nonlinear information transmission between these individuals in local neighbourhoods. As media theorist Eugene Thacker put it:

> The parts are not subservient to the whole—both exist simultaneously and because of each other. [...] [A] swarm does not exist at a local or global level, but at a third level, where multiplicity and relation intersect. (Thacker 2004)

This third level precisely designates a specific adaptive environment, which mediates between external environmental forces and the behavior of swarm individuals.

As a consequence, this chapter seeks to contribute to a more detailed understanding of 'adaptive environments' by exploring the impact of SI—and particularly, the potential impact of swarm robotics (SR)—for architecture. It critically discusses their capability of synchronizing individual movements with influencing environmental forces. The chapter explores how their 'intelligence of movement', or 'logistical intelligence', can be exploited for constructural and building purposes. And it argues that even though the emergent and non-linear capacities of computational SI applications pose intriguing challenges to prevalent architectural paradigms like parametricism (see Schumacher 2009; suckerPUNCH 2010), and although the buzzword SI first was introduced in a paper on collective robotics (Beni and Wang 1993), the transformation into concrete building processes realized by robot collectives is by no means a next step of a linear history towards ever more refined technologies. Swarm Robotics not only pose a set of entirely different hardware and manufacturing problems, but at the same time also lead to adjustments in the conception of dynamic, self-organized design and building processes when these are confronted with the task of constructing the—mostly static—exosceletons of built environments.

The chapter is organized in three sections. The first part critically discusses the theoretical and conceptual entanglements of swarm intelligence and adaptive environments. Finally, both termes allude to a non-trivial hybridity between biological, technological and even ecological traces, terms, and trajectories. The second part provides a critical overview of a number of seminal computational approaches to architecture which derive from the SI mindset and which make use of the adaptability of self-organizing computational agents. These scrutinize novel connections between architecture as a site of material composition and as a site of spatial practices by computer experiments in software environments—be it architectural design tools that generate 'swarm effects' or agent-based models for all sorts of movements and actions of computational agents. The guiding hypothesis—which follows the lines of thought of Oosterhuis or Roland Snooks—is that SI technologies here are primarily used to create *diversity*. Subsequently, the third part of the chapter examines in which ways recent advances in collective robotics lead to further materializations of the adaptive capabilities of swarming that go beyond software applications. It presents three state-of-the-art examples of SR for architectural construction purposes and ventilates some possible benefits as well as a number of principal shortfalls: Although SR—primarily in the form of Unmanned Aerial Systems (UAS), but also as grounded collectives— since several years has developed into a thriving field with a high impact e.g. in logistics, agriculture, or the military, such collective systems seem *principally* rather poorly suited as platforms for architectural building: Besides their limitations in terms of payload capacity, they depend on a working environment which consists of easily identifiable elements, and, at best, shows a lot of regularity in the environment itself (i.e., even surfaces, etc.). If such conditions are not provided, the complexity of using SR for building purposes by far exceeds the costs and means that are needed for other (automated) building technologies. As a consequence, even if there are giant leaps to be expected in automated building and in the use of industrial robots and 3D printers (conceivably with some degree of mobility) (see e.g. Ford 2016, Brynjolfsson and McAfee 2016), the use of autonomous SR building systems *principally* only coheres to very particular environments: Not coincidentally, state-of-the-art papers from this area still resurrect robotic pioneer Rodney Brooks' idea of employing SR for space missions (see Brooks 1989) by focussing on environments where no alternative technologies are at hand, of a similar complex matter, or exhibit little aesthetic requirements. The guiding hypothesis in this third part is that—in contrast to the creation of *diversity* on the SI software level—SR in architecture work best in context with a high degree of standardization and pre-defined modularization, or, on the basis of *regularity*.

1.2 Environmentality

'Adaptive Environments' indicate an exemplary subject matter which connects recent media-theoretical discourses and approaches with architecture and design. Mark Weiser—to refer to him once again—pointed out that "the most profound technologies" of the 21st century "are those that disappear" (Weiser 1991, 94). And Matthew Fuller's seminal publication *Media Ecologies*, at the latest, raised the awareness for the fact that the development of such ubiquitous, mobile, and environmentally embedded media technologies would not only entangle sociosphere and techosphere in unprecedented ways but also emancipate both from humans as their focal point (Fuller 2005). Or, as German media theorists Florian Sprenger and Petra Löffler put it: "In the environment everything is equal—no matter if it is human, animal, plant, or thing" (Löffler and Sprenger 2016: 6). This technological development, says Fuller, can only be understood with reference to ecological modes of description which enable the combination and distinction of heterogeneous elements: These e.g. may include aspects of materiality, technology, biology, sociality, or the political (see Starr 1995). Consequently, it is not a coincidence that media theorists and philosophers like Jennifer Gabrys (2007, 2016), Nigel Thrift (2007), Luciana Parisi (2009, 2013), Mark N. B. Hansen (2014) or Erich Hörl and James Burton (2017) elaborated on these approaches and formulated extensive media-ecological concepts, and that e.g. Petra Löffler and Florian Sprenger suggested to provide some media-historical grounds to this discourse (Löffler and Sprenger 2016).

These authors update a discussion about technical environments for an era of digital cultures which unifies materiality and data transmission. Its conceptual traces, write Löffler and Sprenger, on the one hand lead back to Marshall McLuhan and Neil Postman who, in the 1960s, conceived of media history as a historical succession of media environments—from the alphabet via letterpress printing to electronic media like film, radio and television, and finally to the computer. McLuhan's and Postman's fundamental question always concerned the ways how the appearance of a new medium would transform our structures of perception, thinking, and behavior, and it shows through also in the actual discourse. On the other, it links to Michel Foucault's (2004) conception of the term 'environment' who, in the context of his theory of governmentality, described the redistribution of power relations from defined disciplinary institutions into decentralized environmental agents. But apart form this, the historical strains also point towards ideas from the fields of architecture and urbanism: Patrich Geddes and Lewis Mumford—to name but two protagonists—whisked away the term 'environment' from biology, introduced it to urban studies and cultural theory, and thus connected it with novel areas of knowledge and practice (see Sprenger and Löffler 2016: 9).

If today we speak of technizised or even adaptive environments it is mandatory to not take such terms for granted but to bear in mind the complicated conceptual and theoretical history of their becoming. Sprenger (2018) emphasizes that a profound transformation took places in the discursive trajectory of 'environment' that lead from biology to technical disciplines like architecture. In its early context, that is, in the writings of biologist Herbert Spencer who established the use of the term in the English language in the late 19th century, 'environment' indicated a virtually unchangeable, natural, self-balancing space to which every life form had to adapt to in order to survive. According to Sprenger, during the first decades of the 20th century, this point-blank opposition of environment and man-made modification lost its effective power—to pressing became the urge for controlling environmental factors: Already in the 1920s, early examples extend from ecology, e.g. forestation projects, over the construction of artificial environments as laboratories for the rapidly expanding experimental sciences, to Geddes' approaches to urban planning (see Sprenger 2018).

From there, its conceptual and theoretical history can be continued to the manifold perspectives to understand architecture as a built environment with all sorts of technological and ecological ties—a browse through the headers on arch+or AD cover pages gives a quite appropriate overview. It can be followed as a broader exploration of its environmental sustainability and a critical evaluation of its conceivable contributions to strategies of environmental engineering form a design point of view—as possible answers to the challenges of an actual all-encompassing *environmentality* (see Agrawal 2005). And eventually, it can be extended from Reymer Banham's "well-tempered environments" (1969) to media-technological innovations like digital laboratories, computer simulation environments, or even immersive computer game worlds as well as to those ambient hybrids of architecture, smart materials, and embedded information technology which today wing the steps of investors as sensor-laden smart homes (e.g. Sprenger 2015, 2014), smart cities (e.g. Halpern et al. 2013; Thrift 2014; Kitchin et al. 2017), intelligent workplaces (e.g. Hartkopf et al. 1997), or assistant systems.

The focus on feasible *adaptive* potentials of environments adds a novel twist to the conceptual genealogy of technical environments and exceeds questions of environmental modification: Instead of elements (organisms, things) which *are contained* trying to modificate the containing environments, it now is the containing environment which modifies itself with regard to the necessities of the contained elements (organisms, things). And this twist concurs with an epistemological conversion: McLuhan, in his short text *Message to the Fish* (McLuhan 2001) conveyed that the only thing that fish had no clue of was water—the immediate environment, the containing medium being totally self-evident and taken for granted. He thus alluded to the unreflected adaptation of humans to media environments which he sought to break in furtherance of a critical analysis of his present. Notwithstanding, in the context of adaptive environments this perspective is turned topsy-turvy. Here, it is necessary to explore what the environment knows about its contained elements (organisms, things), how it generates this knowledge, and how it applies this knowledge. Herbert Spencer's organisms which struggled to adapt to an equilibrial environment, as well as later attempts to technically modify, stabilize, or level environmental conditions in the favour of the contained elements are replaced by an environment which adapts to the changeability and the dynamics of its contained elements. Or, to put it another way: Adaptive environments require a theory or a concept of the contained elements to be able to adequately interact with them. And its development becomes all the more demanding the less standardized these elements are or the less predictable they behave. Or, to put it yet differently: The problem of contingency which always complicated the adaption of individuals to environmental forces also works in the opposite direction if technical environments are meant to adapt to the irrationalities and eventualities of contained elements.

In this line of thought, SI and SR can be perceived as exemplary adaptive environments because they approach complex organisation problems by means of artificial *populations* of agents and their behavior in time. The movement paths and vectors of populations, not geometric principles, account for this novel architectural approach. Based on a small number of basic behavioral rules in local neighbourhoods swarms swiftly react a reconfigure themselves dynamically with regard to external disturbances whilst providing the swarm members with a secondary environment that enhances their individual capacities. Architectural design and construction can benefit from the algo-rithmic logics of SI and SR in various ways. First, its mindset extends the possibilities of handling and optimising the complex interplay of various input variables for building processes. It integrates the levels of individual movements of particles (simulated humans, traffic flows, winds, etc.) at the mesoscale of single buildings and at the global level of urbanscapes. Second, the agent collectives—if appropriately tuned—will self-organise in a number of probably interesting or desirable forms over the iterated runs of numerous scenarios, thus transforming the understanding of planning and construction processes. From this change of perspective, architecture becomes based most notably on movements. Moreover, this generation of forms develops in ways that would not be comprehensible without the media-technological means of agent-based computer simulation. Third, it introduces a novel kind of futurology into architec-ture. With computer experiments in ABM software, a great number of different scenarios can be tested and evaluated against each other, offering insight into a variety of different desirable futures. Fourth, this rapid prototyping of possible scenarios in combination with automated procedures of scenario evaluation by evolutionary algorithms introduces a zootechnological and post-humanist element to the design process that can be extended to mass-customized production processes, resulting in a large diversity of forms and shapes in building elements. It thus coalesces more traditional (human) cultural practices of architectural design and construction with novel media technologies. Fifth, the capacity of adding ever more elements to ABM allows for a seamless synthesis of multiple ideas, or for a feedback of opinions by customers or future users during an ongoing design process. And sixth, with SR the prospect of translating such autonomy, flexibility and dynamism to architectural construction is substantiated.

The synthetic character of SI and SR is founded on an underlying algorithmic structure which defines neighbourhoods among all kinds of objects. As an effect, space—in the software and CGI environment of computational swarms and agent-based models as well as in the collective construction procedures of swarm robotics—has no longer to be organised or constituted by a defined geometric grid, but self-generates out of the multiple local interactions of point clouds, particle swarms, or communication signals between robots. SI and swarm robotics act as adaptive environments as they clarify and enable a perspective on space as a computation environment. As Kas Oosterhuis (2006, 14) puts it:

> Taken to the extreme all material is a form of information, and taken even further all information is a form of computation. Thus space computes information. The question to be raised here is: does the space compute or do the people in the space compute? In the context of Swarm Architecture I understand human action in such a way that it must be the space which does the trick. The space is full of more or less active components, many of them communication with each other, many of them interacting with certain intervals, and many of them interacting in real time. [...] How can we look at space with this in mind? Then it is the space itself that behaves and acts, as driven by their programmers and executed by a variety of actors, among them people, but also light bulbs, refrigerators, vacuum cleaners, sofa's, shopping, bookshelves, tables and chairs. They all move or are moved inside a certain space. In the mind of the Swarm Architect, all actors/players behave in relation to each other following a set of simple rules. And it is the space which defines the workspace of the players.

If the main difference which is produced by architecture is the one between inside and outside—as systems theorists from Niklas Luhmann to Dirk Baecker (1990) have claimed—then SI and SR operate as mediators at this exact threshold between inside and outside, at the same time integrating external environmental forces and internal individual forces, and thus processing knowledge of either side.

1.3 Diversity

Swarm Intelligence and Swarm Robotics are entangled from the onset. In 1988, Gerardo Beni and Jing Wang were giving a short presentation on so-called cellular robots—at that time an emerging field of computational methods based on the use of cellular automata—that is, "groups of robots that could work like cells of an organism to assemble more complex parts"—at a NATO robotics conference when in the ensuing discussion they were asked for a buzz word "to describe that sort of 'swarm'." Beni and Wang (1993) took up this suggestion and published their paper with the title *Swarm Intelligence in Cellular Robotic Systems*: A term had been coined which interestingly was first picked up e.g. in fields like biology or in (mathematical) optimization, and in logistics and epidemology (see e.g. Bonabeau et al 1999, Kennedy and Eberhart 1995), transforming the 'cellular robots' and the abstract CA time- and space grids of the 1980s into more flexible ABM. Long before maturing into a technology which was embodied in actual robotic collectives, Beni's and Wang's 'robots' performed their SI in software environments—as computational agents. Nonetheless, the significant principle remained unchanged: "The production of order by disordered action" which appeared to Beni and Wang as the basic—and intriguing— characteristic of swarms (Beni 2008b: 153).

When considering how SI and ABM systems help to treat complex architectural problems, one has to distinguish between two strains of self-organization principles: The one looks at the dynamical generation of (architectural) forms in social insects, the other is occupied with the dynamic movement and adaptive capacities of flocks or schools on the move (like birds or fish). For architectural design, they serve several functions: First, they can be used to produce idea models—that is, inspiring new shapes for further design measures—as an outcome of emergent processes. Such idea models would not have taken on form without the algorithmic logic of SI and ABM (Mammen and Jacob 2008). Second, they can be used to represent the dynamics of existing architectural spaces in a simulation system, facilitating a play with parameters and a testing and evaluation of different scenarios. Third, SI and ABM models from other research fields—for instance, from evacuation studies or pedestrian and traffic simulation (see Helbing 2009 for an overview)—can produce relevant insight which could be integrated in the design processes. And fourth, novel fabrication techniques like mass-customization or 3D printing can be attached to these computational tools which translate the virtual models into material fabric.

The social insects principle relies on a communication structure that uses *stigmergy*, or, more generally, *sematectonic communication* (see Grassé 1959; Bruinsma 1979; Karsai and Pénzes 1993; Bonabeau 1999). This means that the locally defined agents orient themselves not only according to the behavior of a number of neighbours, but also tally traces which the agents place in and read from their environment—like pheromone trails to a food source which produce a positive feedback for following individuals, or of nest structures like honey combs that determine and incite the building of subsequent structures. This distributed organization has been formalized in computer simulation models like *Ant Colony Optimization* (ACO) and initially gave rise to the field of SI (see Bonabeau et al. 1999). In this ABM paradigm, agents collectively transform the incoming information into behavioral patterns and in concrete building structures at the same time.

Here, perception of an environment is transposed from an animal characteristic to an information relation with the aid of a visual interface to make it understandable to the human operator, as media historian Jussi Parikka points out (Parikka 2010: 156). In a seminal publication on SI, Eric Bonabeau, Marco Dorigo and Guy Theraulaz devote a chapter on the computer simulation (CS) of nest building in social wasps. With a three-dimensional Cellular Automaton and carefully evaluated rule sets, they simulated the emergence of a nest architecture which one would find in natural wasps (Bonabeau et al. 1999: 205-252). Stemming from this, computer scientists sought to transform the use of the respective CS technologies from confirming scientific hypotheses to the generative and semi-autonomous development of e.g. *Swarm-driven Idea Models*. Here, the simulation environment works as a virtual testbed for the 'breeding' of complex emergent architectural constructions. In order to result in structures which are somehow suitable for a given architectural problem, the simulators integrate an evolutionary algorithm into the CS which rates the constructional activities of a population of randomly chosen swarms. This consecutively leads to a new population based on the rate-dependent selection of the previous generation of swarms, whilst random changes and recombinations of successful swarms enable the development of unforeseen constructions. In a repetitive process, the CS system yields interesting architectures according to a set of pre-defined evaluation criteria (Mammen and Jacob 2008: 118). Thus, SI enables an integration of architecture into the site-specific environmental context and takes into account aspects of ecological and economic performance of the building (ibid. 2008: 122–124). Whilst one should rather be careful with such tendencies to overemphasize the 'natural integrity' of such outcomes of biologically inspired CS, in terms of a generative approach to the generation of architectural idea models, such *Insect Media* seem to accomplish rather interesting outcomes. However, these are highly dependent on the processually defined boundary conditions of the CS, the design of the learning algorithm which defines the development and 'optimization' of the generation of forms, and not least the expertise of the meta-modeler, the architect.

The second principle in SI is based on the abovementioned movement vectors of flocking individuals defined by local neighbourhoods. Here, the focus lies in the emergence of a dynamic and mutable swarm-space, an intermediate layer between local information processing and collective adaptation to the constantly changing exterior forces of an environmental space. This technique is used for the time-based and dynamic generation of formely unknowable global forms by the non-linear interactions of many mobile individuals. Fueled by sophisticated CGI techniques, ABM softwares were soon embraced by a number of architectural design teams. They transformed creation into merely developing adequate rules which would govern the assembly of components, thus leaving the architect with the role of a meta-designer of self-organizing systems (see e.g. Buus 2006).

Along with other digital techniques such as parametricism (e.g. Schumacher 2009), computational ABM can be networked with digitally controlled production measures. In contrast to traditional building methods, such a 'machine ecology' of file-to-factory mass-cusomization can lead to an endless variety of different building element which are still based on a set of simple rules, and with humans only intervening on a programming meta-level. As an effect, everything is different in absolute size and position, not because of human non-accuracy, but thanks to computational processing of diversity. [...] The driving forces to organize the behavior of the control points of the geometry come from both external and internal forces communicating with the evolution of the 3D model (Oosterhuis 2012).

On the one hand, control thereby is handed to the bottom-up self-organization of non-linear agent systems, on the other it is re-introduced by architects and experts who evaluate the generated forms with respect to certain criteria: "With the centrality of population thinking, the emphasis shifted from both individuals and generalized types to the primary of variation and deviation. [...D]ifference and process become comprehensible and hence controllable" (Parikka 2010: 167).

Roland Snooks, one of the collaborators in an architectural project called *Kokkugia*, explains how ABM methods deal with explicit architectural problems, and how this differs from many of the earlier approaches to digital architecture. *Kokkugia* has been focused on agent-based methodologies [...]. This started as an interest in generative design, not necessarily as a specific interest in computational, algorithmic or scripted work, but as an interest in understanding the emergent nature of public spaces [...] of Melbourne and how we could develop emergent methodologies. That led us to develop swarm systems and multi-agent models (see suckerPUNCH 2010).

But this raises the question of how exactly to define the architectural problem. Due to the non-linear relationality (Thacker 2004) of all objects of a public space, the meta-designers seek to describe all sorts of relations of those objects in simple rules. In this way, the micro-relations of individual agent behavior connect with a meso-scale of giving form to single buildings and to a macro-scale of generative urban planning. With ABM software, as Oosterhuis states, such a system will display real time behavior, and the parameters may change continuously over time. The crucial thing is that comprehensiveness only emerges by running the processes. Using the tentative technologies of SI and ABM in generative architecture thus always seems to be a question of how to shape the bottom-up system behaviors with target functions in a gamified trial-and-error process. Otherwise, reasonable results or idea models would merely be a matter of luck (or patience).

> The challenge for the designer is to find those rules that are effective and which are indeed generating complexity. Some design rules produce death, others proliferate life. Some design rules create boring situations, other rules may generate excitement. You can only find the intriguing rules by testing them, by running the process. (Oosterhuis 2006: 25)

Moreover, instead of working with black boxed modules of commercial architecture software like *Rhino*, *Grashopper* or *Processing* with their respective SI *Boid Libraries* or *Plethora* plugins, people like Snooks advocate the development of open source programs, specific to the respective design intention: "[T]he algorithm should emerge from the architectural problem rather than simply the architecture emerging from the algorithm." (suckerPUNCH 2010).

Broadening this understanding, the collaborators of the *Kokkugia* project describe swarm-based urban planning as a simultaneous process of self-organizing agents which would not any longer result in a single optimum solution or master-plan, but in a flexible near-equilibrium, semi-stable state always teetering on the brink of disequilibrium. This allows the system to remain responsive to changing economic, political and social circumstances. (Leach 2009: 61)—or, in other words, it results in an adaptive environment. In addition, the objective to understand urban dynamics by swarm intelligence systems for *Kokkugia* coalesces with generative measures of their non-linear methodologies to produce shapes of buildings and with the ensuing development of novel fabrication techniques. These could lead to a rethinking of tectonics and form on the basis of ABM (suckerPUNCH 2010). As an effect of SI and ABM models with their focus on moving patterns and dynamic flows, the relationship between locally acting autonomous agents and the material composition of architectural buildings and sites can take on novel operational forms.

These computer simulation systems integrate the effects of spacial practices—that is, the agents' movements—in the material urban fabric, and likewise the constraints imposed on those practices by its (computer-simulated) physicality:

> The task of design therefore would be to anticipate what would have evolved over time from the interaction between inhabitants and city. If we adopt the notion of 'scenario planning' that envisages the potential choreographies of use within a particular space in the city, we can see that in effect the task of design is to 'fast forward' that process of evolution, so that we envisage—in the 'future perfect' sense—the way in which the fabric of the city would have evolved in response to the impulses of human habitation (Leach 2009: 62).

SI and ABM thus can be defined as adaptive technologies which facilitate the apprehension of future states of buildings or urban spaces under varying environmental impacts, carrying the potential to deeply change and enhance the procedures of urban planning. One of their main endowments seems to be the procedural production of diversity—in their use as idea models as well as in combination with the possible mass-customization of building parts involved in construction processes which follow from the computational models.

However, at least two factors have to be paid attention to: First, the smoothness with which some of the most popular SI plugins produce ›appealing architectural forms‹ runs the risk of underestimating effects on rather ›trivial‹ considerations of functionality or tectonics of a resulting structure on the part of the meta-designers. In addition to such digital manierism, a second factor has to be be kept in mind: That is, that such processes of scenario building become as well a part of the reality which they try to model. But in contrast to weather simulations, for instance, the modeled systems—that is, maybe the people using an urban plaza—would certainly react to the scenarios produced by urban planning tools of this kind if those would be on display, say, at a community meeting. Such an interaction of the public with computer simulations that *model* this public would likely add a novel layer of unpredictability to the process.

1.4 Regularity

Whereas Beni's and Wang's paper which coined the term SI lead from cellular robotics right into the realm of computational software applications and ABM, another paper from the same year of 1989 proved more visionary with regard to the development of swarm robotics. At MIT Artificial Intelligence Lab, robotics pioneer Rodney Brooks, together with his working group, was searching for an alternative way to achieve intelligent behavior which contested the cognitivist approaches of GOFAI: Brooks believed that only in relation and interaction with the complexities of a surrounding environment, robots would be capable of developing intelligent behavior. The key term was *embeddedness*, and the conceptual principle was *bottom-up*: Knowledge about the world should rather be computed on-the-run by small robots capable of sensing only those conditions of their environment and react accordingly that were needed to fulfil certain tasks—like, moving around—than by complicated robots with complex artificial brains containing large pre-programmed 'concepts' about the surrounding world. And whilst the MIT Lab more and more began to resemble a zoo crowded by small autonomous robot prototypes—the most popular being *Genghis*, a six-legged ›insect‹ robot without based on a 'subsumption architecture' without a central controller that followed swarm principles internally—Brooks together with Anita M. Flynn pictured the future of and a possible field of application for such machines in a paper boldly entitled *Fast, cheap, and out of Control. A Robot Invasion of the Solar System* (1989: 478):

> Complex systems and complex missions take years of planning and force launches to become incredibly expensive. The longer the planning and the more expensive the mission, the more catastrophic if it fails. The solution has always been to plan better, add redundancy, test thoroughly and use high quality components. Based on our experience in building ground based mobile robots (legged and wheeled) we argue here for cheap, fast missions using large numbers of mass produced simple autonomous robots that are small by today's standards (1 to 2 kg). We argue that the time between mission conception and implementation can be radically reduced, that launch mass can be slashed, that totally autonomous robots can be more reliable than ground controlled robots, and that large numbers of robots can change the tradeoff between reliability of individual components and overall mission success. Lastly, we suggest that within a few years it will be possible at modest cost to invade a planet with millions of tiny robots.

This introduction already compiles almost all ingredients that also today make swarm robotics a compelling approach when it comes to coping with complex demands in unpredictable environmental conditions—its greater robustness, flexibility, reliability, and scalability (see also Brooks et al. 1990). Or, simply put: "[U]sing swarms is the same as 'getting a bunch of small cheap dumb things to do the same job as an expensive smart thing'." (Corner and Lamont 2004: 335). And there is also the economic argument: Small robots can be mass-produced, adding economies of scale, and can be largely constructed from off-the-shelf components. Nevertheless, whilst SI and ABM software applications—thanks to rapidly increasing computing power to calculate the interconnected non-linear behavior of large numbers of agents—began to flourish from the 1990s onwards, swarm robot invasions had been a long time coming (Kube and Zhang 1993). It took more than 15 years until Erol Sahin published the seminal volume *Swarm Robotics* (Sahin 2008), with Gerardo Beni authoring an introduction with the title *From Swarm Intelligence to Swarm Robotics* (Beni 2008a) in which he directly addressed this issue:

> [T]he original application of the term [SI] (to robotic systems) did not grow as fast. One of the reasons is that the swarm intelligent robot is really a very advanced machine and the realization of such a system is a distant goal (but still a good research and engineering problem). Meanwhile, it is already very difficult to make small groups of robots do something useful. (ibid. 2008a: 7)

And even if the volume included reports on pioneering projects like *SWARM-BOTS* (Groß et al. 2006) and *I-SWARM* (Seyfried et al. 2005), the featured discourse remained mostly 'idiosyncratic': It circled around questions of how to engineer functioning robot collectives in the first place whereas the mentioning of concrete application areas was universally rubricated under 'future developments'. This time-lag is—apart from the challenges of engineering working physical systems instead of virtual agents—also due to a changing understanding of SI. In 2000, Sanza Kazadi introduced the term *Swarm Engineering* recognizing that—in contrast to the benefits of emergent effects that are used, for instance, in *Kokuggia*'s computational exper-iments—"the design of predictable, controllable swarms with well-defined global goals and provable minimal conditions" was mandatory in the field of robotics. "To the swarm engineer", he notes, "the important points in the design of a swarm are that the swarm will do precisely what it is designed to do, and that it will do so reliably and on time." (Brambilla et al. 2012, 2, cf. Kazadi 2000). The robots's being out-of-control had to be framed by rigidly determined objectives and behavioral control and—to a comparatively small extend—in some collective robot systems survived in the actual autonomous process of executing the building tasks.

However, the 'distant goal' had been approached rather quickly: In the following the research in collective robotics shows a significant take-off, with today leading to about 1,500 hits for 'swarm robotics' on the *IEEE Xplore* platform alone. Researchers imagined a whole range of possible applications like collective minesweeping or the distributed monitoring of geographic spaces and eco-systems. Swarming elements were imagined to also take on counter measures by self-assembling into blockings against leakages of hazardous materials, thereby being scalable according to the graveness of a situation. The swarm-bots would synchronize with environmental events in space by tracking, anticipating, and level them by self-formation (see e.g. Beni 2008b).

From around 2005 onwards, some strains of research also developed around the operation of swarm robotics for architectural construction (Saidi et al. 2008; Mam-men et al. 2005; Werfel et al. 2006, 2007, 2014; Magnenat et al. 2012; Stroupe et al. 2005; Augugliaro et al. 2013; Mammen et al. 2014; Soleymani et al. 2015; Wawerla et al. 2002; Helm et al. 2012) grounded in the expectation that they not only can […] lead to significant time and cost savings, but their ability to connect digital design data directly to the fabrication process enables the construction of non-standard structures (Willmann et al. 2012: 441).

In addition, at least theoretically, robotic constructive assembly processes are by nature 'additive', they are scalable and can incorporate variation in the assembly to accommodate not only economic and programmatic efficiency, but also complex information about individual elements and their position (Willmann et al. 2012: 446).

And finally, swarm robotics have several advantages compared to already existing platforms: First, unlike common robotic building systems which still are centered around human involvement, swarm robotics could be employed in contexts where a direct human involvement is impractical or too dangerous. Second, swarm robotics overcome the stationary method of common robotic building platforms. Unlike the latter, they are not restricted by the size of the platform, which in common systems have a footprint which must be larger than the final structure. And third, a multi-robot assembly makes use of parallelism and offers error tolerance by substitution, as the sub-tasks can be carried out by any robot of the collective (see Petersen 2016).

Recent research efforts in swarm robotics for architectural building can be roughly subdivived in a four-field matrix containing (1) grounded or (2) aerial robots, which use (3) rigid or (4) amorphous building materials. The typical grounded robot is small, lightweight, and manoeuvrable, equipped with sensors that allow for orientation in the environment and for interaction with other robots and with the building material. Basic challenges for operating such systems are e.g. power supply (battery charging periods), mutual collisions or blockages of robots moving around in a given environment, calculation of shortest paths, and reliable mechanisms for identifying, grabbing, and deploying building materials (see Gerling and von Mammen 2016).

State-of-the-art systems like *marXbot* (Bonani et al. 2010), the *SRoCS* Swarm Robotics Construction System (Allwright et al. 2014), or *TERMES* (Werfel et al. 2014) thereby use highly standarized, rigid building material like cubics or—in case of *TERMES*—blocks specifically designed to meet the robots' manipulators and lifting devices. *TERMES*, which can be perceived as a temporary apex of the scientific field of swarm robotics, is inspired by the decentralized communication structure and collective behavior of termites. The team developed an interaction algorithm for a multi-agent system motivated "by the goal of relatively simple, independent robots with limited capabilities, able to autonomously build a large class of nontrivial structures using a single type of prefabricated building material" (Werfel et al. 2014: 755). After running their algorithm with software agents, the research group implement it in a group of physical robots to test its functioning 'in vivo'. Quite strikingly, *TERMES* commenced to collectively put together the building bricks. Such blocks—as is referred to also in the other seminal research projects—need the capability to adhere to each other or to be mechanically joint, because the use of a secondary material would further complicate the overall process, whilst the robots respectively employ stigmergy as guidance for the exact positioning of the building elements.

However, there are also approaches, which involve amorphous material. Some researchers experimented with sandbags (Napp et al. 2012), whilst others (Napp and Nagpal 2014; Hunt et al. 2014) used amorphous foam to build ramps in uneven terrains, thereby exploiting an advantage of non-rigid materials: The flexibility and thereby the adaptability of the amorphous material vastly facilitated the construction task in that respective environment, whereas their viscosity and expansion introduced imprecision into the construction process (see Gerling and von Mammen 2016). Gerling and von Mammen thus propose a combined process which involves the spread of amorphous materials to even out irregular terrain and the subsequent use of rigid materials "for precise and swift construction" (ibid.). Although, the latter again poses great challenges when it comes to building up tall structures—in this regard, most systems are limited to the range of their lifting devices. *TERMES* however are able to pile their buildings bricks also to temporary ramps which they are able to climb in order to construct taller structures (Petersen et al. 2011, 2014).

In comparison with grounded robots, aerial robots obviously have more freedom to navigate and—with the nowadays favorably employed quadrocopters—also a high degree of precision. They can work dynamically in three dimensions. Although, where the former are most likely to simply stop and shut down if something interferes with its functioning, the latter run the risk of crashing more easily, and thus need a very accurate control for battery charge. Moreover, they are only fitted to transport relatively light loads, which also affects battery size and thus operation time. This disadvantage also remains present in attempts to increase the versatility of amorphous building material by mixing two-component polyurethane to be 'printed' by aerial robots (Hunt et al. 2014).

Nevertheless, UAVs are better suited to build elevated structures (see Gerling and von Mammen 2016; Augugliaro et al. 2013). For instance, the *Aerial Robotics Construction Group* (ARC), a joint research project of two reseach groups at ETH Zurich created a prototype six-meter-tall *Flight Assembled Architecture* tower which contains 1500 foam-brick modules and was assembled by a swarm of autonomous quadrocopters (Willmann et al. 2012: 441-442). As with *TERMES*, the research team emphazised the importance of the 'nature' of a suitable building material:

> The payload of flying vehicles is very much limited, whereas materials with high strength and high density favor the use of ARC [...]. Consequently, this research focuses on the construction of elements, on lightweight material composites and on complex space frame structures [...]. Because the overall shape of these building modules is also determined from aerodynamic considerations, these must be designed according to the specific assembly techniques and building capabilities of the flying machines. The building modules, therefore, must have particular geometrical characteristics so as to meet the required levels of the flying vehicle's complex aerodynamics, and thus, its building performance. The consequence is a design that is never monotonous or repetitive, but rather specific and adaptable to different architectural and aerial characteristics. [...] This ›information‹ logic between dynamic contingencies—such as the requirements of aerial transportation and the physical constraints of production—must be seen as integral. (Willmann et al. 2012: 446-447)

"A design that is never monotonous or repetitive, but rather specific and adaptable"—this perspective certainly can be contradicted. Already the aesthetics of ARC's prototype flight-assembled brick towers and walls, as well as their *SUPERSTUDIO*-like renderings of future megastructures, both prove different (Willmann et al. 2012: 454). Moreover, in the ARC as well as the TERMES example, the autonomy and the adaptive capacities of the robotic swarm collectives are highly integrated with fitting 'environmental interfaces' which on the one hand touch the physicality of the outer environment (e.g. air resistance, irregular surfaces), and on the other the technical specifications of the respective robots (payload, form of building materials, identifiability by building blocks (for instance by RFID tags), sequencing of tasks, etc.). Combined with the necessary reliability in terms of producing satisfying results—that is, the swarm engineering paradigm—it is, as an outcome, little surprising though that most of the contemporary swarm robotic systems—including *TERMES* and *ARC*–execute detailed pre-calculated blueprints. Their adaptivity is the result of a carefully pre-planned system of specifications for standardized building elements.

Thus, statements like the following sound rather lofty if one acknowledges that the respective prototypes still only perform in the artificial environments of laboratories with their radically reduced amount of contingency:

> While it remains to be seen whether ARC will emerge as a viable dynamic building technology, the *Flight Assembled Architecture* prototype successfully illustrates how an ARC approach makes empty airspace tangible to the designer, and addressable by robotic machinery (Willmann et al. 2012: 442).

And furthermore, the abovementioned processes contradict the initial idea of the SI mindset. As swarm robot pioneer Marco Dorigo and his team put it in a paper on their *SRoCS* platform:

> Current implementations of decentralized multi-robot construction systems are limited to the construction of rudimentary structures such as walls and clusters, or rely on the use of a blueprint or external infrastructure for positioning and communication. In unknown environments, the use of blueprints is unattractive as it cannot adapt to the heterogeneities in the environment, such as irregular terrain. Furthermore, the reliance on external infrastructure is also unattractive, as it is unsuitable for rapid deployment in unknown environments. (Allwright et al. 2014: 167)

Their *Swarm Robotics Construction System* avoids the use of a blueprint by enabling the robots to adapt their positioning on visual clues from the environment alone—for instance, they independently identify obstacles or irregularities—and from the building elements which are equipped with 2D bar codes and different lights that indicate their respective status. After positioning the building blocks the robots update the colors of the LEDs on the blocks. Depending on the algorithm in use, these colors can be assigned various meanings, e.g. a particular color can be used to indicate a seed block or a block that has already been placed into the structure, thereby developing the stigmergic building process (see Allwright et al. 2014: 163).

However, in contrast with the sophistication of architectural design and possible mass customization procedures enabled by computational SI application, the physical implementation of collective building processes in swarm robot systems until today remains rather clumsy. Instead of a massively increasing variation of building elements stemming from emergence- and complexity-prone design processes which integrate and calculate a large number of possible agent behaviors, environmental forces, and random fluctiations, swarm robotics is based on careful preparation and pre-planning which—for the most part—eliminates contingency. Working with highly standardized elements and in almost all cases with blueprints or central planning modules, it dimishes the vivid secondary adaptive environments of the computational approaches to mere basic functions, like preventing robots to crash. Hence, the already non-trivial task of constructing reliably functioning robot collectives of larger sizes—see Harvard University's *KILOBOT*-project as a pivotal example which is composed of a stunning 1000 individual robots but comes with a no less dazzlingly slow speed of (re-)arranging collectively (Rubenstein et al. 2012)—is multiplied when it comes to use them as useful construction platform.

For the time being, and compared to already existing (robotic) technologies in architecture, swarm robotics seems to involve rather too much restrictions and disadvantages—for instance in terms of aesthetically and conceptually sophisticated architectural results—and seems to offer rather too few advantages—like being able to autonomously explore terrains and environments which are inaccessible for humans. It is therefore not a coincidence that the *SRoCS* paper leads back to the beginning. Contemplating its possible application area, it is straightforwardly echoing Rodney Brooks's 25-year old vision:

> It is possible that a multi-robot construction system will be a practical solution in the future for building basic infrastructure, such as shelter, rail, and power distribution networks on extraterrestrial planets or moons, prior to the arrival of humans. (Allwright et al. 2014: 158; see also Khoshnevis 2004).

1.5   Conclusion

The chapter demonstrated that swarm intelligence and swarm robotics can be perceived as exemplary adaptive environments. Both approach complex assembling problems by means of self-organizing processes of artificial *populations* of individual agents and their interactional behavior in time. SI and SR thus substitute geometric principles by 'visions of process' as generative forces for architectural design and construction. The emergent and adaptive capacities of swarms on the collective level can be regarded as a mediating layer between exterior influences from the physical environment and the individual actions of swarm members. Understood as a 'secondary environment', swarm systems hence offer multiple benefits for architectural design and construction: First, its mindset integrates the levels of individual movements of particles (simulated humans, traffic flows, winds, etc.) at the mesoscale of single buildings and at the global level of urbanscapes; second, with the capability of rapidly generating diverse scenarios, they can serve as idea generators in the design and construction process; third, this also leads to the integration of futurologic aspects to the design process since computer experiments can direct to previously unknown but desirable outcomes; fourth, such ideas can literally materialize by combining SI design applications with rapid prototyping and mass customization strategies; fifth, such applications can also integrate e.g. customer feedback and can lead to seemless feedback loops over the entire design process; and last but not least, with SR the prospect of translating such autonomy, flexibility and dynamism to architectural construction is substantiated. In a threefold way, the chapter explored the technological history of SI and SR as well as present applications. It thereby discussed in which ways and in which contexts the abovementioned potentials are already utilized.

The first section situated SI and SR as a peculiar form of adaptive environment on a broader conceptual plane which nicely connects the currently burgeoning media-cultural discourse of 'media ecologies' and *environmentality* with the more application-oriented approaches to adaptive environments in architecture. It thereby provided some historical traces of the conceptual transformation from biological and ecological backgrounds to technical environments whose understanding seems mandatory for a comprehensive account of the term 'adaptive environment'.

The second section provided a critical overview of a number of seminal computational approaches to architecture which derive from the SI mindset and which make use of the adaptability of self-organizing computational agents. By distinguishing approaches to self-organization which are oriented at social insects from those which simulate flocks or schools on the move, it also discussed the transformation of the role of the architect into a meta-designer: Using the tentative technologies of SI and ABM in generative architecture thus always seems to be a question of how to shape the bottom-up system behaviors with target functions in a gamified trial-and-error process. One of their main endowments is the procedural production of *diversity*—in their use as idea models as well as in combination with the possible mass-customization of building parts involved in construction processes which follow from the computational models.

And finally, the third section differentiated current developments in SR for architectural building in a four-field matrix of (1) grounded or (2) aerial robots, which use (3) rigid or (4) amorphous building materials. It focused on three state-of-the-art projects, namely the *TERMES* robotic building system of Harvard University, the *SRoCS* Swarm Robotic Construction System, and the *ARC* Aerial Robotics Construction Group of ETH Zurich, and discussed their particular layout and their performance achievements and difficulties. This analysis showed that, unlike SI in architectural design, SR in architectural construction is based on careful preparation and pre-planning which—for the most part—eliminates contingency. Working with highly standardized elements and in almost all cases with blueprints or central planning modules, the secondary adaptive environments of the computational approaches is diminished to mere basic functions—like preventing robots to collide. The question remains whether such robotic building technologies continue to be a highly specialized field for extreme physical environments, which are unsuitable or intractable for traditional methods, or whether they can follow an optimistic 'vision of process' and proliferate into buzzing swarms of rigorous mobile 3D-printers—a vision which would truly be revolutionary for building processes.

Acknowledgements The author wants to express his gratitude to Florian Sprenger (Frankfurt) for discussing parts of his forthcoming monograph on an epistemology of environmental knowledge in a research colloquium at Leuphana University Lüneburg in 2016. He also thanks Sophie Godding and Rahel Schnitter (MECS Lüneburg) for their help with copy-editing this chapter.

References

Agrawal A (2005) Environmentality: technologies of government and the making of subjects. New ecologies for the twenty-first century. Duke University Press, Durham

Allwright M, Bhalla N, El.faham H, Pinciroli C, Anthoun A, Dorigo M (2014) SRoCS: leveraging stigmergy on a multi-robot construction platform for unknown environments, 158–169. In: Dorigo M (ed) Swarm intelligence: 9th international conference, ANTS 2014, Brussels, Belgium, September 10–12, 2014; Proceedings. Lecture Notes in Computer Science, 8667. Springer, Berlin

Augugliaro F, Mirjan A, Gramazio F, Kohler M (2013) Building tensile structures with flying machines. In: IEEE/RSJ international conference on intelligent robots and systems, (IROS) Tokyo, November 3–7, 2013). IEEE, pp 3487–3492

Baecker D (1990) Die Dekonstruktion der Schachtel. Innen und Außen in der Architektur. In Unbeobachtete Welt. Über Kunst und Architektur, ed. by Niklas Luhmann, Frederick D. Bunsen, Dirk Baecker, 83. Haux, Bielefeld

Banham R (1969) The architecture of the well-tempered environment. University of Chicago Press, Chicago

Beni G, Wang J (1993) Swarm intelligence in cellular robotic systems. In: Dario P, Sandini G, Aebischer P (eds) Robots and biological systems: towards a new bionics?. Berlin, Heidelberg: Springer Berlin Heidelberg, 703–12, https://doi.org/10.1007/978-3-642-58069-7_38

Beni G (2008a) From swarm intelligence to swarm robotics. In: Sahin E, Spears WM (eds) Swarm robotics. Springer, New York, pp 3–9

Beni G (2008b) Order by disordered action in swarms. In: Sahin E, Spears WM (eds) Swarm Robotics. Springer, New York, pp 153–172

Bonabeau E (1999) Artificial life 5, 2, Special Issue: Stigmergy

Bonabeau E, Dorigo M, Theraulaz G (1999) Swarm intelligence. From natural to artificial systems. Oxford University Press, New York

Bonani M, Longchamp V, Magnenat S, Rétornaz P, Burnier D, Roulet G, Vaussard F, Bleuler H, Mondada F (2010) The marXbot, a miniature mobile robot opening new perspectives for the collective-robotic research. In: 2010 IEEE/RSJ international conference on intelligent robots and systems, 4187–93, https://doi.org/10.1109/IROS.2010.5649153

Brambilla M, Ferrante E, Birattari M, Dorigo M (2012) Swarm robotics. A review from the swarm engineering perspective. IRIDIA Technical Report No. TR/IRIDIA/2012–014

Brooks RA, Maes P, Mataric MJ, More G (1990) Lunar base construction robots. In: Proceedings IROS'90. IEEE international workshop on intelligent robots and systems' 90. Towards a new frontier of applications. IEEE, 389–392

Brooks RA, Flynn AM (1989) Fast, cheap, and out of control: a robot invasion of the solar system. J Br Interplanet Soc 42:478–485

Bruinsma OH (1979) An analysis of building behaviour of the termite Macrotermes Subhyalinus, Rambur (PhD thesis). Wageningen: Landbouwhoge School, The Netherlands

Brynjolfsson E, McAfee A (2016) The second machine age. work, progress, and prosperity in a time of brilliant technologies.W. W. Norton, New York, London

Buus DP (2006) Constructing human-like architecture with swarm intelligence, http://projekter.a au.dk/projekter/en/studentthesis/constructing-humanlike-architecture-with-swarm-intelligence. Accessed 21 Sept 2017

Corner JJ, Lamont GB Parallel simulation of UAV swarm scenarios. In: Ingalls RG, Rossetti MD, Smith JS, Peters BA (eds) Proceedings of the 2004 winter simulation conference, December 5–8th, 355–363. Winter Simulation Conference Board of Directors

Ford M (2016) The rise of the robots: technology and the threat of mass unenployment. Oneworld, London

Foucault M (2004) Naissance de la biopolitique. Cours au Collège de France (1978–1979). Gallimard, Paris

Friedrich C (2009) Space queries design toolset. Pointcloud-based multi-directional real-time Swarm Architecture design exploration. In: 2009 15th international conference on virtual systems and multimedia: proceedings: VSMM 2009 9–12 September 2009 Vienna. Austria, ed. International Conference on Virtual Systems and MultiMedia. IEEE Computer Society, Los Alamitos, Calif, pp 174–178

Fuller M (2005) Media ecologies. MIT Press, Cambridge

Gabrys J (2007) Automatic sensation: environmental sensors in the digital city. Senses Soc 2(2):189–200

Gabrys J (2016) Program earth. University of Minnesota Press, Minneapolis

Gerling V, von Mammen S (2016) Robotics for self-organised construction. In: 1st international workshop on foundations and applications of Self*Systems (FAS*W). IEEE, 162–167, http://ie eexplore.ieee.org/document/7789462/

Grassé P-P (1959) La reconstruction du nid et les coordinations inter-individuelles chez Bellicositermes natalensis et Cubitermes sp. La théorie de la stigmergie: Essai d'interprétation du comportement des Termites constructeurs. In Insectes Sociaux 6:41–83

Groß R, Bonani M, Mondada F, Dorigo M (2006) Autonomous self-assembly in swarm-bots. IEEE Trans Rob 22(6):1115–1130

Halpern O, LeCavalier J, Calvillo N, Pietsch W (2013) Test-bed urbanism. Public Culture 25(2):272–306

Hansen Mark B N (2014) Feed-forward: on the future of twenty-first-century media. Chicago University Press, Chicago

Hartkopf V, Loftness V, Mahdavi A, Lee S, Shankavaram J (1997) An integrated approach to design and engineering of intelligent buildings—the intelligent workplace at carnegie mellon university. Autom Constr. https://doi.org/10.1016/S0926-5805(97)00019-8

Helm V, Ercan S, Gramazio F, Kohler M (2012) Mobile robotic fabrication on construction sites: DimRob. In: IEEE/RSJ international conference on intelligent robots and systems. IEEE

Hörl E, Burton J (ed) (2017) General ecology: the new ecological paradigm. Theory. Bloomsbury, Bloomsbury Academic, London, Oxford, New York, New Delhi, Sydney

Hunt G, Mitzalis F, Alhinai T, Hooper PA, Kovac M (2014) 3D printing with flying robots. In: IEEE international conference on robotics and automation (ICRA). IEEE, pp 4493–4499

Karsai I, Pénzes Z (1993) Comb building in social wasps: self-organization and stigmergic script. J Theor Biol 161(4):505–525

Kazadi ST (2000) Swarm engineering. Dissertation (PhD), California Institute of Technology

Kelly K (1994) Out of control: the new biology of machines, social systems, and the economic world. Addison-Wesley, Reading, Mass

Kennedy J, Eberhart RC (1995) Particle swarm optimization. In: Proceedings of the IEEE international conference on neural networks. IEEE Service Center, Piscataway, pp 1942–1948

Khoshnevis B (2004) Automated construction by contour crafting—related robotics and information technologies. Autom Constr 13(1):5–19

Kitchin R, McArdle G, Lauriault T (eds) (2017) Data and the city. Routledge, London

Kokuggia (2010) Stigmergic landscape, http://www.kokkugia.com/stigmergic-landscape. Accessed 23 Oct 2017

Kube C Ronald, Zhang H (1993) Collective robotics: from social insects to robots. Adapt Behav 2(2):189–219

Laboratory for Visionary Architecture (2014) Philips Light + Building Pavilion, https://www.l-a-v-a.net/projects/philips-light-building-pavilion-2/. Accessed 23 Oct 2017

Leach N (2009) Swarm urbanism. Architect Des 79:56–63

Löffler P, Sprenger F (ed) (2016) Schwerpunkt Medienökologien. Zeitschrift für Medienwissenschaft, 14

Magnenat S, Philippsen R, Mondada F (2012) Autonomous construction using scarce resources in unknown environments. Autonom Robots 33

von Mammen S, Jacob C, Kókai G (2005) Evolving swarms that build 3d structures. In: CEC 2005, IEEE congress on evolutionary computation. IEEE Press, Edinburgh, 1434–1441. UK

von Mammen S, Jacob C (2008) Swarm-driven idea models—from insect nests to modern architecture. WIT Trans Ecology Environ 113, 117–26 von Mammen S, Tomforde S, Höhner J, Lehner P, Förschner L, Hiemer A, Nicola M, Blickling P (2014) Ocbotics: an organic computing approach to collaborative robotic swarms. In: 2014 IEEE symposium on swarm intelligence. IEEE, pp 1–8

McLuhan M (2001) War and peace in the global village. Gingko Press, Corte Madera, pp 174–190

Napp N, Rappoli OR, Wu JM, Nagpal R (2012) Materials and mechanisms for amorphous robotic construction. In: IEEE/RSJ international conference on intelligent robots and systems. IEEE, pp 4879–4885

Napp N, Nagpal R (2014) Distributed amorphous ramp construction in unstructured environments. Robotica 32(2):279–290

Oosterhuis K (2006) Swarm architecture II. In: Oosterhuis K, Feireiss L (eds) Game, set and match II, On computer games, advanced geometries, and digital technologies. Episode, Rotterdam, pp 14–28

Oosterhuis K (2012) Simply complex, toward a new kind of building. Front Architect Res 1(4): 411–20, https://doi.org/10.1016/j.foar.2012.08.003

Parikka J (2010) Insect media. University of Minnesota Press, Minneapolis

Parisi L (2013) Contagious architecture. MIT Press, Cambridge

Petersen K, Nagpal R, Werfel J (2011) TERMES: an autonomous robotic system for three-dimensional collective construction. In: Durrant-Whyte HF et al (eds) Robotics: science and systems VII. MIT Press, Cambridge, pp 257–264

Petersen K (2016) Collective construction by termite-inspired robots (PhD thesis, 2014). Harvard University, https://dash.harvard.edu/bitstream/handle/1/13068244/Petersen_gsas.harvard.inactive_0084L_11836.pdf?sequence=1. Accessed 28 Nov 2016

Reynolds CW (1987) Flocks, herds, and schools: a distributed behavioral model. Comput Graph 21(4):25–34

Rubenstein M, Ahler C, Nagpal R (2012) Kilobot: a low cost scalable robot system for collective behaviors. In Proceedings of 2012 IEEE international conference on robotics and automation (IRCA 2012). Computer Society Press of the IEEE, Washington, DC

Sahin E (2008) Swarm robotics: from sources of inspiration to domains of application. In: Sahin E, Spears WM (eds) Swarm robotics. Springer, New York, pp 10–20

Saidi KS, O'Brien JB, Lytle AM (2008) Robotics in construction. In: Springer handbook of robotics. Springer, Berlin Heidelberg, pp 1079–1099

Schumacher P (2009) Parametricism: a new global style for architecture and urban design. In: Architectural design 79—Digital Cities, pp 14–23

Seyfried J, Szymanski M, Bender N, Estaña R, Thiel M, Wörn H (2005) The I-SWARM project: intelligent small world autonomous robots for micro-manipulation. In: Şahin E, Spears, WM (eds) Swarm robotics: SAB 2004 international workshop, Santa Monica, CA, USA, July 17, 2004, Revised Selected Papers. Springer, Berlin, Heidelberg, pp 70–83

Soleymani T, Trianni V, Bonani M, Mondada F, Dorigo M (2015) Autonomous construction with compliant building material. In: Advances in intelligent systems and computing 2015. Springer

Sprenger F (2015) Architekturen des environments—Reyner Banham und das Dritte Maschinenzeitalter. In Zeitschrift für Medienwissenschaft 12, 55–67

Sprenger F (2018) (forthcoming) Epistemologien des Umgebens

Sprenger F (2014) Zwischen Umwelt und Milieu—Zur Begriffsgeschichte von Environment in der Evolutionstheorie. In Forum Interdisziplinäre Begriffsgeschichte 3(2)

Starr SL (1995) Ecologies of knowledge: work and politics in science and techology. SUNY Press, Albany

Stroupe A, Huntsberger T, Okon A, Aghazarian H, Robinson M (2005) Behavior-based multi-robot collaboration for autonomous construction tasks. In: 2005 IEEE/RSJ international conference on intelligent robots and systems. IEEE, pp 1495–1500

suckerPUNCH (2010) Interview with roland snooks. 2010 (April 25), http://www.suckerpunchdaily.com/2010/04/25/interview-with-roland-snooks/. Accessed 21 Sept 2017

Thacker, Eugene. 2004. Networks, Swarms, Multitudes. In *CTheory*. http://www.ctheory.net/articles.aspx?id=423. Accessed 21 Sept 2017

Thrift N (2007) From born to made: technology, biology and space. In: Thrift N (ed) Non-representational theory. Routledge, London, New York, pp 153–170

Thrift N (2014) The 'Sentient' city and what it may portend. Big Data Soc 1(1):1–21

Vehlken S (2013) Zootechnologies. Swarming as a cultural technique. Theor Cultu Soc 30(6):110–131

Wawerla J, Sukhatme GS, Mataric MJ (2002) Collective construction with multiple robots. In: IEEE/RSJ international conference on intelligent robots and systems (IROS 2002). IEEE, pp 2696–2701

Wiesenhuetter S, Wilde A, Noenning JR (2016) Swarm intelligence in architectural design. In ICSI 2016: advances in swarm intelligence, pp 3–13

Weiser M (1991) The computer for the 21st century. Sci Am 265(3):94–104

Werfel J, Ingber D, Nagpal R (2007) Collective construction of environmentally-adaptive structures. In: IEEE/RSJ international conference on intelligent robots and systems. IEEE, pp 2345–2352

Werfel J, Bar-Yam Y, RusD, Nagpal R (2006) Distributed construction by mobile robots with enhanced building blocks. In Proceedings 2006 IEEE international conference on robotics and automation, 2006. ICRA 2006. Los Alamitos: IEEE Computer Society Press, pp 2787–2794

Werfel J, Petersen K, Nagpal R (2014) Designing collective behavior in a termite-inspired robot construction team. Science 343(6172):754–758

Willmann J et al (2012) Aerial robotic construction towards a new field of architectural research. Int J Architect Comput 10(3):439–459

Chapter 2
Human-Robot Collaboration and Sensor-Based Robots in Industrial Applications and Construction

Timo Salmi, Jari M. Ahola, Tapio Heikkilä, Pekka Kilpeläinen and Timo Malm

Abstract This paper presents technologies for human-robot collaborative and sensor-based applications for robotics in construction. Principles, safety and control technologies of human-robot collaboration are outlined and sensor-assisted control of industrial robots as well as a dynamic safety system for industrial robots are described in more details. Applicability of sensor-based robotics in building construction and potential of robotics in building construction in general are also evaluated.

2.1 Introduction

2.1.1 Industrial Challenges and Building Construction

Production diversity has been an increasing trend in the manufacturing industry over the past few years. Product life cycles are shrinking, the variety of products is expanding, and production volumes are fluctuating. At the same time, globalization has generated significant pressure to decrease production costs and increase automation. However, the need for flexibility makes automation difficult with well-established technologies. Robot implementations have focused largely on high volume production. As industry seeks new approaches and solutions for flexibility and reconfigurability, solutions enabling robots to work in wholly new applications are to be expected. The building construction includes many work phases that have low-level education standards and difficult working conditions or ergonomics.

T. Salmi (✉) · P. Kilpeläinen
VTT Technical Research Centre of Finland, Espoo, Finland
e-mail: timo.salmi@vtt.fi

J. M. Ahola · T. Heikkilä
VTT Technical Research Centre of Finland, Oulu, Finland

T. Malm
VTT Technical Research Centre of Finland, Tampere, Finland

© Springer International Publishing AG, part of Springer Nature 2018
H. Bier (ed.), *Robotic Building*, Springer Series in Adaptive Environments,
https://doi.org/10.1007/978-3-319-70866-9_2

There is also willingness to restrain the increase of building costs. Robot technology could be seen to have a role in this assignment. However, there are many reasons, why applying robots in building construction seems to be quite challenging. Some reasons to mention:

– Working conditions are difficult for automated machines.
– Working environment is changing while the building progress.
– Most of the buildings are unique.
– The building processes vary; they are different in each case.
– One work phase in one position is quite short; the machine should be transferred at every turn. The work would contain lot of set-ups.
– The accuracy of components and buildings are low, totally in a different level than in metals industry. Different kinds of inaccuracies have to be compensated in several phases.
– The designs are typically inaccurate and imperfect.
– Many work phases require several hands, long reach or continual transferring.

Despite of low-level education standards, flexibility and versatile perceptual ability and ability to react to unforeseeable situation are needed in building construction work. The building construction work has higher requirements for flexibility and adaptability than traditional industrial robot solutions. While advanced robot technology is mostly developed for flexibility needs of industry, it offers possibilities to robot solutions in building construction.

2.1.2 New Possibilities

Potential technologies primarily include advanced sensor technology supported by appropriate software. The most essential sensor technologies are related to machine vision. 2D vision has been the traditional approach, bringing about many successful solutions, and it is becoming a standard option for robot solutions. 3D vision technologies are offering newer possibilities, such as managing in environments that are more difficult and the ability to gather data from 3D surfaces. Sensor technology with related software is a general-use solution for replacing the functions of traditional product-specific mechanical devices. The target is to decrease the amount of product-specific devices to achieve flexibility and the ability to produce different kinds of products with the same equipment and without tedious set-up work.

Another commonly used sensor technology in robotics is force sensors enabling force-controlled processes. Force control is mostly used to adapt a robot to product variation, but also to give better control over robotic processes. Sensor technology opens up completely new possibilities for applications that were unimaginable with traditional technologies. With sensor technologies, robots can identify and locate objects, adapt robot paths to product dimensions, and adapt process parameters to current requirements.

In recent years, human-robot collaboration (HRC) has been of increasing interest in research, and new types of modern safety technology have emerged on the market. These include safety sensors, machine vision based safety systems, laser sensors and safety controllers for robots. The new technology enables flexible fenceless safety systems, common shared work places with humans and robots, and dynamic safety regions alongside a host of other attractive features for human-robot collaboration. The coming of the human robot collaboration is not only a technical issue. New safety regulations and standards have paved the way for new types of solutions. While modern safety technology opens up new possibilities, it also creates new and complex challenges in safety design.

All these technologies have been building opportunities for transferable robotic systems. Transferable systems have also been seen as an attractive edge for enabling configuration/reconfiguration of production systems, lines and workshops. The capacity could be transferred where needed, and investment risks could be reduced. Another potential area for transferability is the ability to process big products with mobile devices, enabling large movements by transferring the devices rather than the product.

2.1.3 Aims and Scope

This paper presents technologies for human-robot collaborative and sensor-based applications for robotics in construction. Section 2 considers principles, safety and control technologies of human-robot collaboration and sensor-assisted control of industrial robots. Section 3 presents a dynamic safety system for industrial robots, and also industrial examples of sensor based robotics. Section 4 discusses applicability of sensor-based robotics in building construction and Sect. 5 gives conclusions by evaluating the potential of robotics in building construction.

2.2 HRC Methods and Principles

2.2.1 Motivation for HRC

Fully automated robotic systems are quite expensive. The robot itself accounts for only a small part of the total investment. A big portion of the system is different devices related to material handling. Unfortunately, they are often highly product-specific and offer little flexibility. Some parts of the process are often difficult to automate, meaning that the devices tend to be expensive. The advantages of robots are their speed, accuracy, tirelessness and force. Robots can repeat monotonous movements accurately, untiringly and without breaks, even with heavy loads. The need for flexibility make systems more complex and it is quite often difficult to get reasonable repayment periods for flexible robot investments.

Human workers, by comparison, are intelligent, creative, highly flexible and able to adapt to new situations and variations in products and environments. A human can work with very simple tools at a workstation that needs little investment. The reverse side is limited effectiveness, the need for breaks, and costs per hour. Fully manual work is not competitive for many products in high-cost countries.

The aim of HRC is to combine the best features of both human and robot. HRC can occur at different levels, whenever a robot and human are in close co-operation and possibly in contact with each other. The idea is to enhance the flexibility of the human and the speed, accuracy and tirelessness of the robot in repetitive work. Some work phases are very difficult to automate; an attractive alternative is to keep them manual. HRC leans more toward light investment, with high flexibility but good efficiency. Traditionally human-robot collaboration has been used seldom due to safety issues and in most cases robots and humans having to be separated by fences and safety devices. Now, new safety technology is opening up completely new possibilities for such operations.

The use of a fenceless robot system is not only for close collaboration. In some cases, the fences feel uncomfortable, making both the layout and movements around production lines tricky. Fenceless robots are considered easier to integrate into production lines with fewer changes than those required for traditional robot systems. Collaborative functions also facilitate monitoring and the resolution of disturbances.

All robot systems require some kind of human attention at different stages, such as robot programming, system set-up, material handling and clearing of disturbances. Real collaboration may involve a human and a robot sharing a workspace.

The most likely dangers related to robots include collision of a moving robot with a human, risk of compression, flying objects and sparks. VTT carried out an analysis of robot accidents before 2006 and found that most severe accidents were compression cases. Very seldom was a robot collision the cause of a severe accident (Malm et al. 2010). This could change, however, once collaborative robotics become more widespread.

2.2.2   What Is Expected from Safety Technology

Safety devices should be designed in such a way that failure does not cause danger. Any failure should halt the robot's function safely. Safety sensors apply the continuous signal principle, whereby an object is detected when the signal is missing or low. Currently, there is one exception to the rule: laser scanners. These can be applied for safety purposes in a relatively clean environment, but dirt or moisture could reduce the detection range to negligible. Depending on the safety level (SIL/PL/Type/Category), failures of the sensor lead to a safety function. This principle ensures that the sensor is either operational or a safety function is triggered (e.g. stop or failure alarm).

2.2.3   Safety Restrictions to Human Robot Co-operation

The conventional solution to safeguarding an industrial robot system is to keep operator and an automatically running robot away from each other. Already the first European robot standard (EN 775:1992) described the safeguarding principle. Basically, this means that in an automated run, the moving robot stops (protective stop removes servo power) before a human can touch it. An exception to this requirement is an enabling device (unofficially called "dead man's switch"), that allows access to the robot working area. Then a reduced speed (<250 mm/s) is applied, and the device stops the robot if its button is released or pushed over the tipping point. Since entering the robot area typically involves cutting the servo power, restarting must be done outside the robot area. The restarting process is slow and does not enable sound collaboration between a person and a robot.

Designers have long dreamt of fenceless robot cells where humans and robots can work safely together. Collaboration modes and safety requirements between human and robot were first described in ISO 10218-1:2006. The collaboration modes are:

- Monitored stopping: the robot stops when a person enters and restarts when the person leaves the robot area
- Hand guiding: The robot is guided by hand by pushing or with a handle
- Speed and position monitoring: robot impact is prevented by controlling speed and separation distance
- Power and force limiting: harmful impacts are prevented by controlling the robot force and power.

Safety-rated monitored stopping stops the robot but keeps the servo power running, allowing humans to enter the robot work area safely. If a robot fails to stand still, a protective stop is launched and power is cut off. Restarting can be automated and it enables sound collaboration with the robot.

Power and force limiting became allowed in 2006 with the possibility to apply inherently safe forces and power (150 N, 80 W, near tool center point). However, the force values were removed from existing harmonized robot standards (ISO 10218-1:2011), (ISO 10218-2:2011).

Currently, standard specification (not harmonized) ISO/TS 15066:2016 describes in greater detail the forces and pressures associated with injury and pain. Limit values are listed by human body part (ISO/TS 15066:2016). The contact point is no longer limited to the tool center point (TCP) as it was in earlier standards. From a safety point of view this is more accurate, but for robot system integrators it is difficult to measure and prove acceptable force values. However, power and force control enable deeper collaboration with the robot than do other accepted collaboration modes.

It is essential that safety functions are realized according to ISO 13849-1 PL d (performance level) and Category 3 requirements. This means that the probability of dangerous failure is low ($<10^{-6}$), dangerous failures are identified (diagnostic coverage > 60%), and there is enough redundancy (duplicated according to category 3). If the safety function requirements are not fulfilled, the safety function cannot be trusted and additional measures are needed. Typical safety functions are safety-rated monitored stop, safety speed, limited work area and limited force. They are applied to minimize or prevent impact by a robot. One possibility is to estimate the severity of an impact and, if no severe injury is possible (small robots), then lower the PL requirement according to the results of risk assessment. Actual additional measures may be related to e.g. sensors, fences, the stopping function, or the position of possible contact points. If a heavy-duty industrial robot cannot match the PL requirements, then, typically, protective stop (servo power off) needs to be applied to ensure safety when a person is approaching the automatically running robot.

2.2.4   New Types of Safety Technology

One representative of the new safety technology is a robot safety controller—a doubled virtual controller that can be used for safety functions to ensure safe operation. Any deviation between the virtual and real controller triggers an emergency stop. A safety controller allows software-based safety functions to be used, like area boundaries designed in a 3D-virtual environment, speed restrictions, and tool direction limitations. It also allows for a safety-rated monitored stop without shutting down the robot. One primary objective of the controller is to fit robots into smaller spaces. For example, it allows several acceptable areas to be designated for the robot that can be switched on according to input signals. The system makes it possible for the robot to have different functional modes that can be used in human-robot interaction in different situations.

A safety laser scanner is one of the most popular new types of flexible and programmable safety devices on the market. It is an optical sensor that scans the environment in a 2D fan shape with an infrared light, measuring the distance to detectable obstacles. The scanning range can be up to 270 degrees and the measuring range around 4–7 m depending on the scanner. Several warning or emergency fields can be configured within the scanning zone using a PC, and active areas can be switched around according to signal inputs. The device can be mounted horizontally or vertically. The disadvantage of horizontal mounting is that it does not detect the hands of an approaching human. Vertical mounting prevents the use of different safety areas, only the boarder of each area can be detected.

Pilz SafetyEye is one of the most advanced 3D safety sensors currently available. The system operates using three cameras assembled at ceiling height, 1.5–7.5 m above the area. The system compares the current view with the defined view and reacts when differences are detected. Several warning and emergency areas can be observed simultaneously in 3D. The resolution depends on the camera height and size of the detected area. The sensor meets the performance level d requirements for controllers (SFS-EN ISO 13849-1 2015). According to tests carried out at VTT, the system determination and programming are fairly clear. Although the system has very versatile safety functions, its drawbacks include the somewhat high investment cost, limited reaction time, and sensitivity to changes in light and other disturbances such as shadows, smoke and sparks.

Inherently safe collaborative robots are a fairly new type of robots. An earlier standard (ISO 10218-1: 2006 Robots for industrial environments—Safety requirements—Part 1: Robot) paved the way for workplaces containing both humans and robots. The standard set the limit for robot speed at 250 mm/s and forces that the robot could initiate against persons at 150 N (static) and 80 W (impact). The more recent Technical Specification ISO/TS 15066, published in early 2016, specifies the force limits more exactly; force and pressure are now limited according to which part of the body the contact is directed at.

Over the last few years, the commercial availability of collaborative robots has risen sharply, with the advent of prototypes and the continuing emergence of new companies and models. One of the main features of collaboration robots is their light structure. Sharp edges have been largely eliminated and surfaces rounded or even softened. Robot structures have been designed to avoid compressive gaps. Maximum robot speed, power and forces can now be defined, and the robots themselves are equipped with a wide range of collision-detecting functions. All these features are designed to minimize the effects of possible contact with humans, forces and pressures. The first inherently safe robots were very slow and light. Now, several 10-kg workload collaborative robots are available, and a prototype with over 100 kg load have been presented.

Universal Robotics is one the pioneers in collaborative robotics. It offers three sizes of 6-axis robot, thee are UR3, UR5 and UR10, with payloads of 3, 5 and 10 kg respectively. The robots have very light arms with ranges of 850–1 350 mm. The repeatability is ±0.1 mm and maximum TCP speed only 1 m/s. Performance is not at the same level as that of traditional industrial robots, but it has other advantages. Programming by guidance is possible, the user interface exceptionally friendly, and the programming easy to learn.

Kuka's LWR iiwa is one of the best-known HRC-compatible robots. This 7 axis, light arm robot has a payload of 7 or 14 kg and reach of 800 or 820 mm. Each axis has a force sensor for use in safety functions and force-controlled operations or as part of the human-robot interface. The robot's movements can also be guided manually. (Kuka 2017) The LBR iiwa controller uses Java technology for sequence programming to achieve modularity, openness and simplicity, but is not very user-friendly. The robot offers several advanced and versatile functions, but is quite highly priced.

ABB's YuMi, launched in 2015, is a fairly new arrival on the market. According to ABB, the "inherently safe" design allows the YuMi to work alongside humans while reducing risk to acceptable safety levels. The robot has two 7-axis lightweight arms with softened and rounded surfaces. It has a 0.5 kg payload and good accuracy of 0.02 mm. YuMi is designed to work alongside humans on assembly lines. Its dimensions are fairly similar to those of a manual worker. It also has integrated servo-grippers with optional built-in cameras. The robot controller is also integrated into the robot body (ABB Robotics 2017).

2.2.5 Principles of Human-Robot Collaboration

Collaborative operation may include one or more of the following methods, according to ISO/TS 15066:

– Safety-rated monitored stop
– Hand guiding
– Speed and separation monitoring
– Power and force limiting.

The traditional approach has been to shut down the robot whenever a human needed to enter the robot area, for example to load parts into magazines. Restarting the robot could be unreasonably difficult. A safety-rated monitored stop allows an operator enter the robot area safely when the function is activated. The robot stops moving but is not shut down, and it can resume its activity once the operator has left the area.

In the hand guiding method, an operator uses a hand-operated device located near the robot end-effector to transmit motion commands to the robot. Before an operator enters the collaborative area, the robot must have stopped safely. The guiding device is activated manually and robot movements are also controlled manually. The robot speed has to be safe limited.

The speed and separation monitoring method allows the robot and an operator to work simultaneously in a collaborative workspace. The risk of a collision between human and robot is reduced by keeping the protective distance between them at all times. The distance between a human and a robot is monitored by safety technology and the robot speed varies depending on the distance. If the distance falls below the protective separation distance, the robot system stops.

According to current standards (ISO 13855 2010 and ISO/TS 15066), the protective distance for a collaborative robot is calculated with the formula: $S = K_r (T_S + T_r) + S_r + C_{Tol} + K_h (T_s + T_r + T_b) + Z_d + Z_r$, where:

S_r is the robot stopping distance, which can be over the working radius of the robot;

C_{Tol} is a factor based on the recognition abilities; here the factor $8 \times (d-14 \text{ mm})$ is used, e.g. when $d = 40 \text{ mm} => C_{Tol} = 208 \text{ mm}$;

T_S, T_r are the reaction times of the sensor and robot, and T_b is the robot stopping time;

K_r is the approaching speed of the robot;

K_h is the approaching speed of the human operator (1.6 – 2.0 m/s);

$Z_d + Z_r$ are the position uncertainty of the operator and robot from their position measurement systems.

The calculation is based on the assumption that a robot and a human can approach each other before the safety sensors react and the robot stops. The protective distance is strongly dependent on the robot's speed, because it affects the robot's breaking distance and breaking time. Also, all devices have reaction times that should be taken into account. The protective distance and safety devices ensure that the robot always stops before coming into contact with an operator.

The above methods can be realized with traditional robots equipped with safety-related control features, but the power and force limiting method requires robot systems specifically designed for this type of operation. With this method, human and robot can be in contact with each other, but potential hazards are kept below threshold limits of force and pressure set according to different parts of the human body. A collision can occur during an intended work task, or during incidental contact in unforeseeable situations or as a result of failure. Risk reduction is achieved in an inherently safe robot through passive and active safety design. Passive design includes increasing the contact surface area with e.g. rounded corners and smooth surfaces. It can include padding to absorb energy and extend energy transfer time and to minimize moving masses. Active safety design includes robot safety-related control functions like limiting forces, velocities of moving parts, momentum and power. It also includes workspace limiting or soft-axis functions and proximity or contact detection sensing to reduce forces. Potential contact situations should be identified, analyzed and, if possible avoided, and the contact forces and pressures kept below limit values. The risk assessment must include all devices in an application, including robot tools, work pieces, etc. A combination of risk reduction methods may be necessary, and possible additional collaboration methods would be needed, such as safeguarding. Objects with sharp edges etc. cannot come into contact with humans.

2.2.6 Safety Challenges of Transferable Robots

Transferable robotic systems set special requirements for safety system design:

– The system set-up after transfer should be made fast and easy.
– Investments in safety should be cost-efficient: safety devices should be relatively cheap and the safety system preferably wholly integrated into the transferable platform.
– The safety system should allow efficient utilization of space.

The safety system is preferred to be suitable for many different situations and production environments without extensive changes to the work environment. Transferable robotic systems can be divided into two groups: freely transferable robotic systems and systems that have multiple prepared regular workplaces (e.g. for machine tending) between which the robot cell is transferred according to production needs. Safety of the transferable robotic cell is considerably easier to achieve with prepared bases than with freely transferable systems. However, many problems still arise that are not encountered in stationary robot cells (Salmi et al. 2013).

Optical safety sensors are best suited to integration in transferable systems. Mechanical fences can be transferred but are inconvenient to rebuild after each transfer. One challenge related to optical safety sensors is the long protective distance. A safety fence can be built directly at the maximum reach distance of the robot, but when using an optical safety sensor there is free access to the robot area. The stopping time of a bigger robot may be around 1 s, during which a human can walk 1.6 m. This leads to long protective distances and poor utilization of space. The robot speed should be lowered to reduce breaking time and drop the protective stop. If the robot area is to be less than the robot's maximum reach, it is possible to restrict the robot. Then it is necessary to ascertain whether the function prevents the robot from exiting the restricted area or whether is activated when the robot actually exits. In the latter scenario, the breaking distance must be added to the protective distance. It is possible to use space efficiently with a safety-related robot area limitation, be there should be a concomitant speed limitation (Salmi et al. 2013a).

When transferring a robot between prepared positions, previously customized setups can be used for the safety sensors. With a freely transferable robot, the situation is more complex. Preferably, the work done after transferal should always be new. New risk assessments and setups for the sensors should be done and the safety functions tested. It is possible to integrate safety scanners into the transferable robot platform, but the robot itself creates a shadow, as may many necessary auxiliary materials and equipment. Adjustment of safety zones to permanent obstacles needs setups for the safety sensors. In the end, adjustable mechanical fences or transferable light curtains are not much more uncomfortable to use. However, the safety design of a freely transferable robotic system varies from case to case, and to date there is no known satisfactory solution.

2.2.7 Methods for Co-operative Robot Control

Here we discuss the methods for co-operative robot control and specifically human-robot collaboration in physical contact. The most common methods to implement co-operative robot control are to monitor the electric current consumed by the joint servo motors, to measure the joint torques or to directly measure the external forces and torques affecting the robot's end-effector. The first two methods are typically applied to lightweight robot arms with relatively low (typically below 10 kg) weight-carrying capacity such as Universal Robots, KUKA's Iiwa, ABB's Yumi, and Rethink Robotics' Sawyer. However, in construction and industrial applications the handled parts are often much heavier than 10 kg, even exceeding 100 kg, and the only viable option is to use medium-sized or heavy-duty industrial robots with non-backdrivable gears in the joints. Because of the high friction in the gears, the monitoring of motor current or measuring of joint torques is unsuitable for implementing co-operative control. Instead, for high-payload robots, cooperative control is typically implemented by mounting one or more force/torque sensors in the robot's end-effector to measure the external forces affecting the robot, and then to use some external force controller to modulate the robot's position based on the measured forces and torques.

A well-proven method for robustly controlling the contact forces between a position-controlled robot and the environment is the impedance control approach (Vukobratovic et al. 2009; Ahola et al. 2015; Brunete et al. 2016). The main idea of

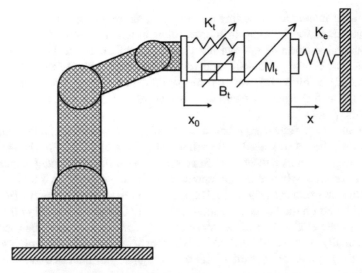

Fig. 2.1 Schematic of an impedance-controlled robot in contact with the environment; K_t is the stiffness of the robot, Bt is the damping effect of the robot, x_0 is the location of the robot (flange), M_t is the inertia of the robot, K_e is the stiffness of the environment and x is the location of the contact between the root and the environment

impedance control is to measure the external forces and torques acting on the robot's mounting flange, and to the make the robot's end-effector follow the second-order dynamics of the mass-spring damper system, as shown in Fig. 2.1.

The end-effector of the impedance-controlled robot follows the equation of translational motion, where applied contact force depends increments of position (stiffness), velocity (damping) and acceleration (mass/inertia) of the end effector. The rotational motion of the robot's end-effector follows the equation and rotational motion, where applied contact torque depends increments of position of the rotation angles (stiffness), velocity of the rotation angles (damping) and acceleration of the rotation angles (mass/inertia) of the end effector. In robotics applications, the impedance parameters may be set separately for the x-, y- and z-axes and for rotations around the x-, y- and z-axes of the TCP or specific compliance frame.

The challenge of robot impedance control is how to ensure contact stability between the robot and a varying environment without too much damping. By adjusting the damping parameter, the contact motions can be effectively stabilized, but excessive damping causes the robot to react sluggishly to external forces, which in the worst case might make it inappropriate for human-robot collaboration.

Specifically, in the hand-guiding mode a viable option is to utilize two force torque sensors, the first sensor in the robot's mounting flange and the second sensor in the guiding handle (Brunete et al. 2016; Ahola et al. 2017). The first impedance controller can be designed for hard contact between the robot and the environment, and the second impedance controller for soft contact between the robot and a human operator. Thus, much lower damping parameters may be used for controlling the human-robot interaction, enabling a highly responsive co-operative robot system. The lower limit for damping in human-robot interaction depends fundamentally on the communication delays and limit frequency of the robot. Unfortunately, the mechanical links of high-payload robots are heavy and thus also the limit frequency is low (typically below 2 Hz), which means significant damping is needed to stabilize the contacts and the human operator will always feel some resistive forces when guiding the robot.

When using two force-torque sensors in hand guiding, the coordinate transformation between the sensors need to be calibrated (Ahola et al. 2017). The calibration of the pose parameters between the force torque sensors is important, especially in collaborative tasks where a human operator guides the robot with a tool in contact with a stiff environment (Fig. 2.2). The robot motion against a rigid object can be restricted based on the forces measured with the F/T sensor mounted in the robot's flange (Brunete et al. 2016). However, without accurate calibration the robot could be accidentally overloaded because an F/T sensor mounted in the guiding handle supplies the impedance controller with much higher gain than one mounted in the robot's flange.

2.2.8   Sensor-Assisted Control

In addition to collaborative hand-guiding applications, the impedance control approach may be used during automatic path execution for controlling the contact forces between the robot and the environment. Adjustment of robot paths based on measured process forces may be utilized in material-removing manufacturing processes such as grinding and polishing. The force controller maintains the pressure between the tool and surface within a predefined range and compensates for the wear of the tool as well as deviations in the work object dimensions.

In addition to force control, the robot paths may be adjusted based on optical object-locating sensors such as CCD Cameras, 2D laserprofilers and 3D cameras. These sensors may be used for recognizing and locating work objects in the robot's workspace before path execution (Heikkila et al. 2010; Ahola et al. 2016; Ahola and Heikkilä 2017) or for adjusting the robot paths during path execution, such as in welding seam tracking. The adjustment of robot paths during path execution requires sensor data processing in real-time and sending path corrections to the robot controller via an interface with low latency. Whichever the method of application, the object-locating sensors need to be properly calibrated with respect to the robot (Heikkila et al. 2014). In this context, calibration means determining the full 6-degrees-of-freedom coordinate transformation from the sensor's coordinate system to the robot's. The sensors fixed to the environment are calibrated with respect to the robot's base coordinate system and the sensors mounted in the robot's end-effector are calibrated with respect to the robot's mounting flange (Tsai and Lenz 1989).

The stationary 3D camera can be calibrated with respect to the robot as shown in Fig. 2.3a). In the calibration sequence, the robot places the calibration plate in a set of

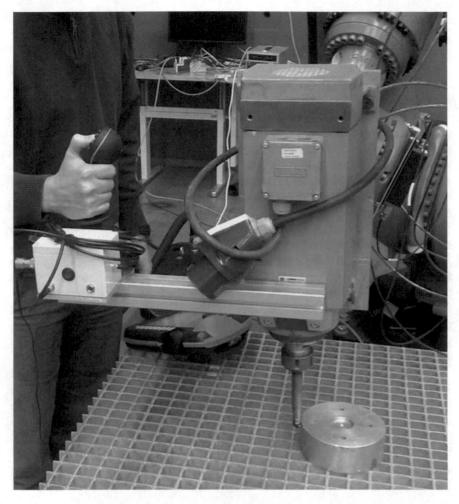

Fig. 2.2 The robot's tool is guided in contact around a stiff object (Ahola et al. 2017)

poses and a point cloud is grabbed from each pose. The robot's flange poses and the points belonging to the calibration plate are recorded and supplied to the calibration function, which gives a least mean squared estimate for the coordinate transformation from the robot base coordinate system to the camera coordinate system. The segmented points belonging to planar surfaces are shown in Fig. 2.3b). The purple points are measured from the table plane and the blue points from the calibration plate.

2D machine vision has been the standard solution for flexible object localization in industrial applications for decades. During the past ten years, several industrial 3D sensors and sensor systems have become commercially available, but 3D machine vision applications are still quite rare in industry. One apparent reason for this is the lack of general-purpose commercial software for processing the 3D sensor data. Typically, sensor manufacturers provide only the drivers to grab the 3D point clouds from the sensor, but not the software for filtering and analyzing the images. Some exceptions exist, such as SICK PLB (Sick AG, Waldkirch, Germany), SICK PLR (Sick AG, Waldkirch, Germany) and Pick-It (Pick-It N.V., Gaston, Belgium), but their analyzing software may not be compatible with other vendors' sensors.

VTT Technical Research Centre of Finland has developed general purpose sensor non-dependent object recognition and locating software for varying-resolution 3D sensors (Ahola et al. 2016; Ahola and Heikkilä 2017). VTT's software is built on the Robot Operating System (ROS), which is an open-source middleware for handling communication between multiple computing processes called ROS-nodes. Currently, VTT's object recognition and locating software can process data from many 3D sensors, such as the SICK LMS100 (SICK AG, Waldkirch, Germany) tilting scanner, SICK Ruler E1200 (SICK AG, Waldkirch, Germany), Basler Time-of-Flight (ToF) camera (Basler AG, Ahrensburg, Germany), Zivid (Zivid Labs A.S., Oslo Norway), Kinect Version 1 and 2 (Microsoft, Redmond, Washington USA) and Primesense Carmine 1.09 (Apple, Cupertino, California, USA). Although 3D machine vision requires more computational processing than 2D machine vision, it has some significant advantages. It enables estimation of the full 6-degrees-of-freedom object pose from a single image, correlating CAD models with measured data, estimating parameters of 3D surface patches, and determining the dimensions of objects more accurately.

The localization results of a cubic target object measured with a Zivid 3D camera is shown in Fig. 2.4a), and with a Basler time-of-flight 3D camera in Fig. 2.4b). The Zivid 3D camera and Basler ToF camera have obvious differences in resolution and accuracy, and hence the sensors require very different settings for processing the point clouds. The Zivid, which is based on structured light, provides a 1920 × 1200 image with 0.1 mm depth resolution, while the Basler ToF camera provides a 640 × 480 image with 10 mm depth resolution. In comparison, the measuring range of the Basler ToF is up to 13 m, while the optimum range of the Zivid is 0.6–1.1 m. As seen

(a)   **(b)**

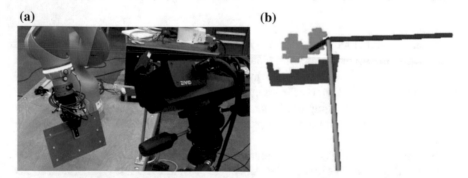

Fig. 2.3 **a** The Zivid 3D camera is being calibrated with respect to the robot's base coordinate system; **b** segmented planar surface patches

(a)   (b)

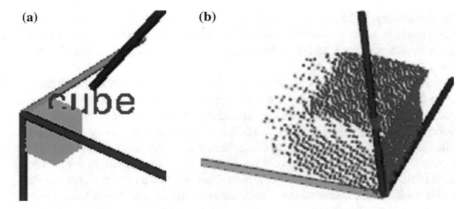

Fig. 2.4 a Cubic target object recognized and located from a point cloud measured with a Zivid 3D camera and **b** Basler time-of-flight 3D camera

in Fig. 2.4b), the point cloud cluster includes both "flying" pixels, i.e. outliers, and distorted points, making the cube facets seem somewhat curved. The outliers and distorted points are characteristic of ToF sensors, making their deployment for 3D machine vision applications challenging. Although Zivid provides a very dense and accurate point cloud requiring hardly any filtering, the challenge is to find optimal camera settings applicable for various materials, colors and glossy surfaces.

2.3 Applications

2.3.1 Dynamic Safety System

The objective was to develop an advanced safety arrangement enabling collaboration between a human and large industrial robot in a shared workspace. The aim was to build the system according to current safety regulations. The other key principle was to avoid unnecessary emergency stops after human attention.

The safety of human-robot cooperation must comply with standards ISO 10218-1, ISO 10218-2 and ISO/TS 15066. According to the standards, two alternatives can be considered: The robot has to be stopped before a human can touch it, or the collision forces must be very limited. Since we were dealing with a large robot with a lot of moving mass, we contemplated an HRC with speed and separation monitoring. The protective distance ensures that the robot is stopped before a human can touch it in any situation.

One challenge was that calculations of protective distance lead to very long distances and excessive use of space. A safety-related function of restricting the robot working area makes it possible to restrict the robot workspace. One problem is that the safety controller is often a monitoring parallel robot controller that reacts when the robot exceeds determined limitations. This means that the robot breaking distance has to be taken into account. Given that the breaking distance of a bigger robot at maximum speed can be 2 m, area restrictions are meaningless if speeds are not limited (Salmi et al. 2013b).

As noted in the case of a transferable robot system, reduction of the robot velocity is an effective way to reduce a robot's stopping time and especially its breaking distance. Each stopping situation is unique and dependent on the robot speed, load, position and path. Also, all robot installations may have a variety of situations. The stopping situation should be handled through a worst-case scenario that is very difficult to determine. This can be done for the current robot within the set speed limit (Salmi et al. 2013a). The developed system is called a dynamic safety system, and it consists of two subsystems. The primary system can be based on any human detection sensor. It detects the movements of a human worker, changes the mode of the safety system, and reduces the robot speed according to the predicted worker position (Fig. 2.5). The primary system aims to avoid activation of the safety system (Salmi et al. 2016).

The secondary system is a certified safety system based on optical safety sensors and a robot safety controller. The safety system has several setups or modes, having speed and area limits for the robot and for the safety sensor emergency area based on related protective distance. The safety system mode is changed according to the human detection system. The dynamic safety system can bring the robot to a monitored stop without executing an emergency stop. Also, automatic recovery to normal speed is enabled when a human has left the danger area. The human worker is informed of the system status via informative graphical displays (Salmi et al. 2016).

The dynamic safety system allows a human to walk around the robot workspace safely. The robot adjusts its speed according to the distance to the human. When a human approaches the robot it slows down, if necessary, even to a controlled stop. As soon the human draws away, the robot can start again and increase its speed. A safety configurator has been developed to help make configurations for the dynamic safety system. In each case the maximum speed, maximum workload and allowed work area of the robot must be set. The system uses a gathered database of robot stopping distances at different speeds and workloads. In the software tool, the user can place different sensors and robots from the library to a layout representing the dynamic environment. The scanning areas of the sensors are shown with the created safety areas, which are generated automatically based on the given robot speed and workload. (Several) different safety configurations are created related to different parameters. It is also possible to define different robot work areas in order to make the system more dynamic. The configuring tool shows the necessary separation distance, thus all created configurations comply with the safety regulations. The tool can also import and export safety configurations to safety devices (Salmi et al. 2016).

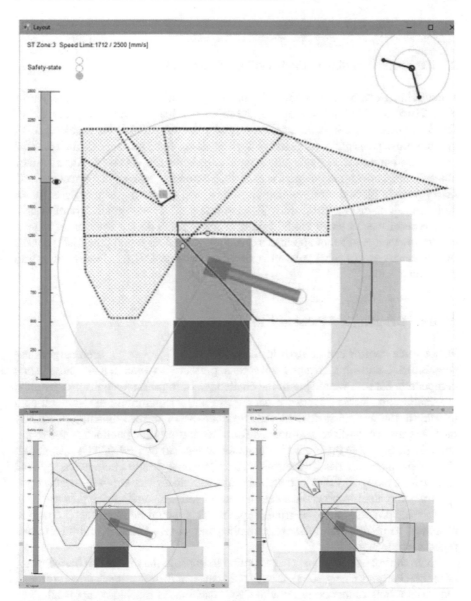

Fig. 2.5 Principle and demonstration of the dynamic safety system: When a human approaches the robot, the allowed robot area (green lines), robot speed and safety zones (the red area) of the laser scanner change dynamically

2.3.2   Industrial Examples

2.3.2.1   Hand Guiding in the Handling of Large Parts

Hand guiding can be used in the HRC handling of heavy parts and objects (Ahola et al. 2015). In this kind of scenario, the robot is a load carrier with 6 degrees of freedom guided by a human operator. In general, collaborative hand-guided robot systems are expensive for use as lifting devices alone. To be profitable in an industrial setting, the hand-guiding application should exploit the robot's positioning accuracy and automatic operation. Potential uses include assembly, where the robot could hold heavy parts in place during welding or bolting, and palletizing, where parts and/or pallet locations change frequently. The advantage of a hand-guided robot system is its ease of use, as small modifications of robot paths do not necessarily require expert knowledge of robot application programming. This could potentially cut the maintenance costs of robot cells and make the investment worthwhile for industrial end-users.

2.3.2.2   Grinding and Polishing

Robot force control can be used in material removal processes, such as grinding (Fig. 2.6) and polishing, to adjust the surface pressure between tool and surface and compensate for tool wear. One of the challenges of impedance-controlled grinding is selecting applicable impedance parameters and programming the nominal path to achieve the desired surface quality, for example in surface finishing. Currently there are no standardized instructions for selecting process parameters for force-controlled robots, as there are for computer numerical control (CNC) machining. The robot integrator has to determine the applicable process parameters by trial and error, making it difficult to estimate the technical risks and total costs of the system beforehand. A potential solution in short series manufacturing is to use hand guiding for programming the grinding paths. This would allow expert production personnel to participate in the programming and more efficiently find the optimal process parameters.

Force control has previously been applied to improve the machining performance of robots (Pan and Zhang 2008). However, because of their low mechanical stiffness, industrial robots cannot compete with CNC machines in machining applications in general. This is because industrial robots are kinematic chains with long linkages, whereas CNC machines are parallel kinematic machines designed for the highest possible mechanical rigidity. Low mechanical stiffness means not only poorer machining accuracy, but also lower feed and material-removal rates than with CNC machines.

Fig. 2.6 Grinding with a force-controlled robot

2.4   Applicability in Building Construction

Prefabrication has been identified as one of the most promising targets for robot automation in construction (Vähä et al. 2013). Several tasks in building construction, especially in prefabrication, are quite similar to industrial applications. Large element dimensions are a challenge for industrial robots (reach of the robot not sufficient, extra axis needed for moving robots or building elements), as is the heavy weight of building elements especially when these are of concrete. Tasks involving only moderate forces and accuracies have the most potential for robotization. This chapter introduces methods for the robotized manufacturing of molds for concrete components, with a practical casting example.

2.4.1   Robotized Machining of Molds for Concrete Castings

In construction, a wide range of component shapes and concrete structures, such as concrete foundations and decorations, are produced using molds. There is a growing need to produce molds for concrete elements with specific curved or even arbitrary shapes, like in the restoration of historical artifacts or new artistic designs.

Conventional formwork favors straight lines and sharp corners, resulting in more or less simply formed concrete blocks and structures. Wooden materials (wood, plywood) are those most often used for concrete molds, but reusable metal frames are becoming more popular. Flexible materials like fabric formwork involving the use of structural membranes have been applied as a building technology (Veenendaal et al. 2011), resulting in forms exhibiting curvature and highly finished surfaces, which is unusual for concrete structures. However, for specific or even arbitrary shapes this is not sufficient.

Heikkilä et al. (Heikkilä et al. ISARC 15) have applied digital manufacturing by developing a robotized method for manufacturing individual concrete elements with specific shapes, using the robotized preparation of molds from hardened sand. Digital 3D models of the target objects form the basis from which mold machining paths for the robot are created stepwise as follows:

- Mold design: preparation of the mold geometry (with AutoDesk Inventor®)
- Tool path programming: creation of machining tool paths (with ABB RobotStudio® Machining PowerPac)
- Post-processing of NC paths: conversion of the tool paths to robot paths (with ABB RobotStudio® Machining PowerPac)
- Simulation: simulation of the path execution (with ABB RobotStudio®)
- Post-processing of simulated robot programs: conversion of the robot paths to KUKA-compliant form (with a proprietary post-processor on a Windows PC).

2.4.2   Mold Design

A 3D digital model of the target object forms the basis for the mold design. In principle, there can be any kinds of curved surfaces in the CAD model data, taken e.g. from construction design data such as BIM (Vähä et al. 2013). Another option is to acquire the 3D digital model, for example for renovation targets, by scanning the object with a laser profiler sensor. The resulting 3D point cloud can be transformed to a 3D surface model for the mold design. Figure 2.7 illustrates a 3D digital model, and a comparable mold model prepared with a CAD system (AutoDesk Inventor®) by cutting out the digital face from a mold preform block model. The result is a negative 3D image showing the targeted mold surface profile.

(a)   **(b)**

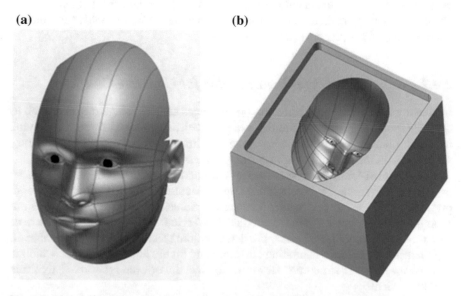

Fig. 2.7   Digital 3D model of the target shape (face) (**a**) and its negative 3D image (**b**)

2.4.3   Robot Programming

The robot programming for mold machining consists of several phases, starting with tool path programming. This is done using a block model of the mold preform and the target mold profile with the negative profile as inputs to the NC programming tool (ABB RobotStudio® Machining PowerPac (RobotStudio Machining PowerPac 2014)). The paths are created by correlating the source geometry, i.e. the mold preform and mold target surface profiles. Other machining parameters, like the machining path types, approach directions, and overlapping are also given. The resulting paths are internally converted to RAPID language, ready to be used in the ABB RobotStudio® simulator tool or in ABB robots. The post-processed paths from ABB RobotStudio® Machining PowerPac are simulated in the RobotStudio® to check the reachability of the path points, joint constraints as well as axis configurations with the initial values for the mold fixture position and tool transformations. Finally, the generated machining paths (with roughly 10000 path points) are converted into the target robot language, like KUKA KRL, at the same time dividing the paths into smaller path segments with a maximum size of 500 path points per segment. This is done using a proprietary converter tool SW running on a MS Windows PC.

2.4.4   Task Execution of Robotic Machining

Before running the machining programs with the robot, the actual location for the work object has to be defined in the robot coordinate system. Also the tool correction, i.e. the exact definition for the TCP, must be defined by 6 or 7 parameters from the robot flange: x, y, z, a, b, c for KUKA robots with Euler angles as rotations, or x, y, z, q1, q2, q3, q4 for ABB robots with unit quaternions as rotations.

(a)   **(b)**

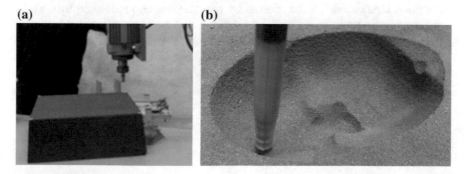

Fig. 2.8 Mold preform **a** and machined shape **b**

The robot is equipped with a milling tool spindle in the robot wrist, and because the spindle is rather heavy (here Perske AC spindle, 10 kW, 400 V, 18A, max. 17720 rpm, weight ca. 50 kg), machining requires a strong and rigid robot to move the tool accurately, in this case a KUKA KR150 or KUKA KR120 R2500 Pro. The mold preform is introduced to the machining stand in the robot cell, and calibrated values of the mold preform location and TCP are fed into the machining program of the robot. The mold preform is composed of hardened sand as shown in Fig. 2.8a. The robot machines the final shape of the mold with a tool of hardened steel alloy, spherical head, R 10 mm. During machining (Fig. 2.8b), the pull-out loose sand material is removed manually with a vacuum cleaner. Running the machining path at about 10 000 points takes around 30 min.

2.4.5   Collaborative Sensor Guidance for Robotized Machining of Large Molds

The workspace of a robot always has limitations, thus machining of the mold does not always occur at one stationary location. The mold may need to be moved from one place to another before machining can proceed. To avoid tedious teaching of the mold location manually several times during multiphase machining, a mold localizing sensor system facilitates the process. Heikkilä et al. (Heikkilä et al. IASTED Robo 10 2010) have introduced a collaborative (or an interactive) low-cost sensor system for this purpose.

In the sensor-based robotized method for mold machining, the process is extended with sensor programming and collaborative sensor measurements. Using the mold localizing sensor system includes two extensions to the robotized mold manufacturing procedure:

– Programming the Collaborative Sensor System: pick up reference surfaces from the mold CAD model, to be marked by the operator and measured by a stereo camera system (with MeshLab®);
– Collaborative sensor task execution: guided by the collaborative sensor program (showing the surfaces to be marked), the robot operator marks the reference surfaces with marker tags, e.g. paper stickers (with an operator PC and a proprietary application program) and the sensor system provides the robot program with the part coordinates.

2.4.6   Collaborative Sensor Programming and Localizing Measurements

For collaborative localization, the robot operator needs to know how the mold (preform or partially processed) should be presented to the sensor system for localization. For this, the mold designer—or alternatively a production planner—examines the mold CAD model and selects the target surfaces (features) to be shown and visualized for the robot operator. This takes place by picking surface points from the reference surfaces and specifying the geometric primitive type of surface (planar, cylindrical, spherical, and conical currently supported). The system then computes the surface parameters, specifying the location and orientation of the surface in the mold coordinate system and the initial position of the mold preform, and finally extends the interactive sensor program and related mathematical model of an object-localizing algorithm (for more details on the algorithms, see Heikkila et al. 2010).

The geometric reference features are parametrized as follows:

- Planar: a point in the plane pp and the normal vector of the plane np.
- Spherical: center point ps and radius rs.
- Cylindrical: center axis with a point in the axis pc and direction vector sc, and the radius rc.
- Conical: center axis with a point in the axis pc and direction vector sc, tip point pt and opening angle a.

The localization is based on detecting the marker points from the programmed reference surfaces. This is carried out with two cameras, where the 3D point is derived from the measured markers in the image planes of the cameras based on pin hole camera models and photogrammetric stereo. Reference surfaces must be specified for all the models of the multiphase machining. Figure 2.9 illustrates mold models for a machining procedure with four phases. A test system for mold machining with collaborative multiphase localization is illustrated in Fig. 2.10. Views from the sensor system (single camera view) and collaborative localization with markers affixed to the mold preform and partially machined mold surfaces are illustrated in Fig. 2.11.

(a)   **(b)**   **(c)**   **(d)**

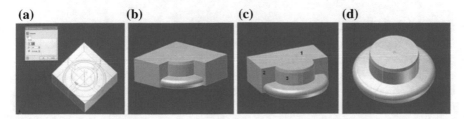

Fig. 2.9 Mold preform **a**, partially machined mold **b** and **c**, and final mold shape **d**. Reference features marked on the partially machined mold model (**c**)

Fig. 2.10 Setup of the mold machining robot system with collaborative localizing sensors (cameras inside the orange circles)

(a)   **(b)**

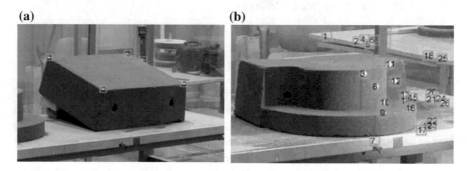

Fig. 2.11 Collaborative localization: views from one camera with markers detected on the mold preform (**a**) and partially machine mold (**b**)

2.4.7   Casting of Concrete Objects

Once the mold is complete, before casting it with concrete its surface must be finalized with a coating to prevent the concrete from adhering to it. Spackle has been found to be a viable option (Heikkilä et al. 2015). After casting, some spackle may remain on the surface of the object. This must be removed with e.g. a water spray. Several demonstration shapes of robot-machined molds have been used in casting experiments. The experiments clearly show that robotized mold machining enables specific shapes of building components to be automatically produced with nearly free variation of curved shapes, even when using a single milling tool with a very large end radius. Using more tools with a smaller end radius can add greater detail to the final shape.

2.5 Conclusions

Here we discuss the challenges and possibilities of construction robotics, including within the context of prefabrication and onsite construction operations.

Prefabrication is a very common construction method for building both small houses and apartment complexes. Prefabricated concrete or wooden elements such as walls, ceilings or floors are fabricated in the factory using series production with automated machines. Fully furnished rooms, such as bathrooms or kitchens, can also be manufactured within the limits of transportability. Element prefabrication is analogous to car manufacturing, which involves assembly, spot welding and painting of car bodies. Several phases still exist that are done fully manually by human workers (e.g. installing insulation). Many manual tasks could be done with robots, but they are so short that investing in robots for the purpose would not be cost-effective. In many cases, the robot system must be able to execute multiple tasks to achieve a sufficient exploitation rate to justify the investment. The technology exists; for instance, an industrial robot could be equipped with automatic tool-changing capability, but the rise in costs and complexity of the robotic system would defeat the purpose of the investment.

Since prefabricated elements have large dimensions, extra axes are needed for moving the robot or element to achieve sufficient reach of the robot. The carrying capacity of the robot also needs to be considered. An industrial robot is best suited to repetitive tasks requiring moderate force, such as installing thermal insulation, stacking wooden parts, nailing, screwing parts together, or painting. The internal logistics of the factory must also be considered, since robot cells need parts, such as precut timber, to be delivered punctually to the production cell.

One challenge is that prefabricated elements are highly customized products. In principle, all the necessary information about them is provided by the building information model (BIM), but there are still challenges implementing it into automatic fabrication. A human worker can work with imperfect information and solve issues as they arise, but it can be challenging for robot systems. Additional sensors, such as a camera system or force sensor, could be used to increase the flexibility of the robotic system. They could help the robot pick parts from a pallet in arbitrary orientations, check the correctness of the parts, adapt to their dimensional variations, find notches or edges when assembling them, and finally check the quality of the work.

Construction sites are a far more challenging environment for robots, which are still a rarity there. Portability is one of the main issues; a robot system must be easy and quick to transfer from one location to another to achieve a high utilization rate. In general, industrial robots are not designed to be portable but can be made so by attaching them to a moving platform. One option is to develop special equipment that can be dismantled and moved easily and that performs specific tasks very efficiently. Automatic or semi-automatic robotic devices could become work partners for human workers, carrying loads or performing simple repetitive tasks such as painting or grinding.

Nonetheless, even relatively simple tasks may require a degree of perception and decision making. For example, a painter manually painting a wall checks the quality of the work constantly, and can decide whether more paint is needed in certain spots. This can be challenging to mimic with sensors and software, but advances in machine learning and artificial intelligence (AI) hold promise for this area.

Existing construction methods were originally developed with human workers in mind, and involve numerous small tasks that are challenging to automate. One approach could be to make changes to construction processes and the structures of buildings so they are easier to manufacture with robots and automation. A hot topic in recent years in construction automation has been the 3D printing of houses. It is an example of a construction method originating from a totally different background that is based entirely on digital design and automation. It is still too early to say whether the 3D printing of houses will evolve into a widely adopted building method, and thus become the first big commercial success story of automation in building construction.

Robot safety is another challenge in building construction. If the robot system can be separated from humans, the solution is clear. In most cases, however, it would need to be transferred easily in changing construction environments. Fully satisfying safety solutions for this purpose are elusive. A sensor-based safety solution would require a new risk assessment and setup after each transfer, which takes time and expertise. Transferable mechanical safety fences might be the most practical in the end. As the development of sensor-based safety systems moves toward automatic configuration, new possibilities may open up to managing speed and distance in HRC. Hand-guided robots would be the most practical approach in several construction tasks, including material handling, lifting, and transferring and joining phases.

References

ABB Robotics (2017) Brochure: YuMi® creating an automated future together. You and me. 14.8.2017, ABB, abb.com/robotics

Ahola J, Heikkilä T (2017) Object recognition and pose estimation based on combined use of projection histograms and surface fitting. In: Proceedings of the ASME 2017 international design engineering technical conferences and computers and information in engineering conference IDETC/CIE 2017, 6–9 August 2017, Cleveland, Ohio, USA. The American Society of Mechanical Engineers ASME, 7 p

Ahola JM, Heikkilä T, Koskinen J, Seppälä T, Tamminen T (2016) A configurable CAD-based object recognition system for varying resolution 3D-sensors. In: 2016 12th IEEE/ASME international conference on mechatronic and embedded systems and applications (MESA), Auckland, pp 1–6. https://doi.org/10.1109/mesa.2016.7587121

Ahola JM, Koskinen J, Seppälä T, Heikkilä T (2015) Development of Impedance Control for Human/Robot Interactive Handling of Heavy Parts and Loads. In: ASME. International design engineering technical conferences and computers and information in engineering conference, vol 9. 2015 ASME/IEEE international conference on mechatronic and embedded systems and applications: V009T07A066. https://doi.org/10.1115/detc2015-47045

Ahola JM, Seppälä T, Koskinen J, Heikkilä T (2017) Calibration of the pose parameters between coupled 6-axis F/T sensors in robotics applications. In: Robotics and autonomous systems, vol

89, pp 1–8, ISSN 0921-8890, https://doi.org/10.1016/j.robot.2016.11.020, (http://www.scienced irect.com/science/article/pii/S0921889015302785)

Brunete A, Mateo C, Gambao E, Hernando M, Koskinen J, Ahola JM, Seppälä T, Heikkilä T (2016) User-friendly task level programming based on an online walk-through teaching approach. Ind Robot Int J 43(2):153–163. https://doi.org/10.1108/IR-05-2015-0103

Heikkilä T, Ahola JM, Koskinen J, Seppälä T (2014) Calibration procedures for object locating sensors in flexible robotized machining. In: 2014 IEEE/ASME 10th international conference on mechatronic and embedded systems and applications (MESA), Senigallia, pp 1–6. https://doi.or g/10.1109/mesa.2014.6935567

Heikkilä T, Ahola JM, Viljamaa E, Järviluoma M (2010) An interactive 3D sensor system and its programming for target localizing in robotics applications. In: The IASTED international confer- ence on robotics. Robo 2010, Phuket, 24–26 Nov 2010 Proceedings of the IASTED international conference robotics. Robo 2010. IASTED (2010), pp 89–96

Heikkilä T, Vähä P, Seppälä TA (2015) Robotized method for manufacturing individual concrete elements with specific shapes. In: International symposium on automation and robotics in con- struction 2015 (ISARC 2015), June 6–8, 2015, Oulu, Finland, 6 p

ISO 10218-1 (2011) Robots and robotic devices-Safety requirements for industrial robots - Part 1: Robot systems and integration, 43 p

ISO 10218-2 (2011) Robots and robotic devices-Safety requirements for industrial robots - Part 2: Robots, 72 p

ISO 13855 (2010) Safety of machinery. Positioning of safeguards with respect to the approach speeds of parts of the human body, 40 p

ISO/TS 15066 (2016) Robots and robotic devices-Safety requirements for Industrial robots — Collaborative operation, 33 p

Kuka (2017) Sensitive robotics_LBR iiwa brochure. 18 s. 04/ 2017, Kuka Roboter Gmbh. Kuka.com

Malm T, Viitaniemi J, Latokartano J, Lind S, Venho-Ahonen O, Schabel J (2010) Safety of interactive robotics—learning from accidents. Int J Soc Robot 2:221–227. https://doi.org/10.1007/s12369- 010-0057-8

RobotStudio® Machining PowerPac (2014) Increased engineering efficiency within machining applications. Data sheet, ABB 2014, 2 p

Salmi T, Marstio I, Malm T, Montonen J (2016) Advanced safety solutions for human-robot- cooperation. In: 47th International symposium on robotics, ISR 2016, 21–22 June 2016, Munich, Germany, Proceedings. mechanical engineering industry association (VDMA); Information Tech- nology Society (ITG) within VDE (2016), pp 610–615

Salmi T, Väätäinen O, Malm T, Montonen J, Marstio I (2013a) Meeting new challenges and pos- sibilities with modern robot safety technologies. In: 5th international conference on changeable, agile, reconfigurable and virtual production, CARV 2013, 6–9 October 2013, Munich, Germany

Salmi T, Väätäinen O, Malm T, Montonen J, Marstio I (2013b) Enabling Manufacturing Competi- tiveness and Economic Sustainability Springer, pp 183–188. https://doi.org/10.1007/978-3-319- 02054-9_31

SFS-EN ISO 13849-1 (2015) Safety of machinery-Safety-related parts of control systems – Part 1: General principles for design. Finnish Standards Association SFS, 193 p

Tsai RY, Lenz RK (1989) A new technique for fully autonomous and efficient 3D robotics hand/eye calibration. IEEE Trans Robot Autom 5(3):345–358

Veenendaal D, West M, Block P (2011) History and overview of fabric formwork: using fabrics for concrete casting. 2011 Ernst & Sohn Verlag für Architektur und Technische Wissenschaften GmbH & Co. KG, Berlin, Structural Concrete 12, No. 3, pp 164–177

Vukobratovic M, Surdilovic D, Ekalo Y, Katic D (2009) Dynamics and robust control of robot- environment interaction. In: New frontiers in robotics, vol 2. World Scientific Publishing Co. Pte. Ltd, 638 p

Vähä P, Heikkilä T, Kilpeläinen P, Järviluoma M, Heikkilä R (2013) Survey on automation of the building construction and building products industry. VTT TECHNOLOGY 109, Kuopio 2013, ISBN 978-951-38-8031-6, p 82

Zengxi P, Hui Z (2008) Robotic machining from programming to process control. In: 2008 7th world congress on intelligent control and automation, Chongqing, pp 553–558. http://ieeexplor e.ieee.org/stamp/stamp.jsp?tp=&arnumber=4594434&isnumber=4592780

Chapter 3
Emancipating Architecture: From Fixed Systems of Control to Participatory Structures

Kevin Clement, Jiang Lai, Yusuke Obuchi, Jun Sato, Deborah Lopez and Hadin Charbel

Abstract Automation revolutionized not only the processes and outputs of manufacturing, but also fundamentally changed the way in which humans participated in the act of making. The result has been a shift from human-centric design and production processes to a techno-centric paradigm. Instead of defining what is commonly construed as a human-machine dichotomy, this chapter examines, how architectural fabrication can be reconceptualized by changing the roles of the different intertwining agents that contribute to the production of physical architectures through the, precedents, and a case study. While robotic production processes often seek to create controlled, efficient outputs, this chapter explores the use digital feedback processes to proactively integrate mechatronic devices, material inconsistencies, and human intuition by weaving them into a network that creates optimized structures through time. While the context, form, and use of the structures may change, each output is clearly identifiable as a part of the same underlying system.

3.1 Introduction

Participating in construction processes has traditionally required a high degree of skill and craftsmanship. Technology has often been viewed as a medium that can minimize human mistakes in the fabrication process and produce material uniformity. However, recent research has begun to question the current roles of the designer, material, and maker in the production of architecture, conceptualizing the final output as an embodiment of a collective effort between each role. As the differences between the real and the digital become indistinguishable, so too is the relationship between these different parties. Increasingly, all three entities can be seen as active agents in a generative fabrication process. The role of the designer can be viewed as one who creates systems that allow for the individual agency of both the material and the maker to be expressed both locally and globally at architectural scale. This chapter

K. Clement · J. Lai · Y. Obuchi (✉) · J. Sato · D. Lopez · H. Charbel
Advanced Design Studies, University of Tokyo, Tokyo, Japan
e-mail: obuchi@arch.t.u-tokyo.ac.jp

© Springer International Publishing AG, part of Springer Nature 2018
H. Bier (ed.), *Robotic Building*, Springer Series in Adaptive Environments,
https://doi.org/10.1007/978-3-319-70866-9_3

53

provides an in-depth analysis of a research project, titled "DRAWN", which seeks to utilize the potential of both human mistakes—inputs—and material properties to generate architecture. These actors are organized via a time-based, digital feedback loop, which proactively seeks to organize the myriad forces that constitute the act of making, linking them together to create a generative production logic that is highly differentiated and physically optimized. The final structures seek not to minimize local differences, but instead to emphasize the traces of the individual hands that have fabricated the structure.

DRAWN is built by combing the use of a 3D-pen with an assembly logic that allows anyone to take part in the fabrication process. This chapter is divided into three sections that address different aspects of this project. Part I contextualizes the work by providing a survey of contemporary research related to additive manufacturing processes. Part II breaks the project down into five separate research categories and explains how each serves to reinforce the participatory aspect of these productions. Part III explains how the research categories were combined at architectural scale to produce two case study projects. The first work, referred hereafter as "Case Study I", measured approximately 2.2 × 2.2 × 1.6 m and was built as a freestanding structure at Ozone Gallery in Shinjuku, Tokyo. The next produced work, titled "Case Study II", was built at the Centre Pompidou in Paris, France as a hanging structure. Finally, Part IV reflects on these results and speculates on potential research trajectories. This chapter is not intended as a technical overview of the project, but rather as a half-way point, which reflects on the past successes and failures of the project thus far. Through these works, the research seeks to redefine the role of the human, the digital, and the material, so as to shift the impetus of architecture from top-down planning, to creative agency.

3.1.1 From Automation to Participation: A Survey of Contemporary Trends in Additive Manufacturing

Advances in design methodologies, material studies, and computational tooling continue to have transformative effects in additive manufacturing, altering both the materialized outputs and the different ways in which they are produced. Rooted in a desire to develop non-standard and unique one-off's, digital fabrication has sought to move past the combinatory possibilities of mass produced parts. This section examines evolutionary trends between different modes of design-to-production, demonstrating a shift away from a vocabulary of the smooth, seamless, and optimized, and entering a new paradigm of the rough, evolved and inclusive. Although these changes are related to technological developments tied to automation, material explorations and digital feedback loops, they have had more subtle implications related to how we perceive of and deal with issues of control.

Initially, additive manufacturing focused on the technological developments of 3D printing at architectural scale by using cement-based materials and CNC processes. This method was first proposed in the latter half of the 90s (Pegna 1997) and has since been widely researched, tested, and proven successful in generating structurally stable outputs. Developed under three different methods known as D-Shape, Contour Crafting (CC) and Concrete Printing (CP) (Lim et al. 2012), the process uses principles of horizontal layering, wherein the nozzle location, speed and material dry-times are precisely calibrated to effectively produce a physical replica of a digital model. This technique, though seminal and fully automated, is generally non-adaptive as the tool path is often hardwired and material expectation predetermined. Therefore, a high degree of control is necessary within both the production setting and the variables (namely the material and nozzle coordinates). Though this method does not allow for deviation from an initial target geometry, works such as the Grotto (Dillenburger et al. 2013) shift the focus of research from material production towards generative and recursive computational design methods. Highly intricate, ornamental outputs are digitally created, thus by-passing the limited nature of fabrication as a deterministic process and giving agency to computational methods that accept the physical limitations of the printer itself. While geometrically rich, the building process is based on a conventional stacking logic of large discrete 3D printed blocks that are read as a whole when assembled, thus maintaining a distinction between design and fabrication.

A continuation of this 3D-printing process utilizes multi-axis industrial robots, thus allowing for increased dexterity, as well as reintroducing anthropomorphic processes back into digital fabrication techniques. Rather than using horizontal contouring, this method is based on three-dimensional spatial tool paths. As a result, the risk of error is increased because the phase changing material is no longer supported by a horizontal datum. For this reason, new strategies for surrendering control while achieving overall coherence are necessitated. One example that explores this issue can be seen in the Voxel Chairs (Retsin and Garcia 2016), where discretized assembly logics are implemented into a continuous printing process. In this instance, the potential discrepancies are localized into smaller voxel zones, allowing inconsistencies to be ultimately subsumed by the whole. Although control is still maintained within the spatial coordination of the nozzle, the imprecision caused by the material behavior is effectively nullified.

Between the above processes is a second method, which found its beginnings in early form finding experiments by Frei Otto. Through the study of self-organizational material aptitude, some materials show a rigorous association between form and structure in biomimetic emergent models of architectural production (DePaola 2012). Here, design preference takes a secondary role to inherent material performance logics. One such example that has evolved this approach through digital tooling can be found in the pavilion researches conducted at the IDC in Stuttgart, whereby morphological complexity and performative capacity from material constituents are explored computationally, providing highly integrated architectural solutions that are able to synthesize material, form and assembly (Menges 2018). By combining the repeatability afforded by the digital with the studied stochastic behavior within phase changing materials, new organizational logics can be uncovered rather than imposed. One such example can be seen in the emergent patterns that result from plotting molten wax into a tank of cool water, which embeds structural tendencies

3.2 Methods

Continuing in the tradition of participatory additive manufacturing, DRAWN seeks to create an adaptable system that is capable of absorbing change through the interplay of architecture's constitutive fabrication processes. Material is organized into a structural system that is built with a hand-held tool; as humans make the structure, a digital feedback system, still in development, is used to guide makers in the creation of adaptable forms. Traditional notions of labor are challenged, wherein play, chance, and "mistakes" become catalysts in the production process. For this reason, while the overall geometry and aesthetic of each structure is consistent and identifiable, the drawing skill and desire of the individuals who fabricate the structures manifests itself locally; when one examines the final project, they can see that different areas of the structure were built by different people. In addition, due to the unique skills of each maker, and because multiple people can make the project simultaneously, the global geometry will be differentiated from the initial intent of the designer. These differences are encouraged and conceptualized as projective acts. Utilizing this system to build structures will result in the creation of different built outputs, even when the initial input is unchanged. This desire runs counter to the tendencies of industrial production processes, which seek to homogenize outputs as much as possible, minimizing difference in favor of consistency.

The architectural system can be divided into five parts as follows: (1) Material—PLA, a thermoplastic in a state of constant, slow deformation, non-hazardous and recyclable; (2) Tool—an ergonomic mechanism that deposits this material in controllable amounts; (3) Organizational Logic—a system for organizing the material; (4) Structure—a method to test the workings of the system; (5) Digital/Real Feedback Loop—a tracking and guidance system capable of locating the position of the material in real-time, analyzing these locations in relation to the "ideal" digital model, and then updating this model to reflect the realities of the fabrication process. This chapter does not elaborate on the technical aspects of each of these categories. Rather, an explanation is given of the theory and intentions behind each part, our findings are explained, and areas that require further research are speculated upon.

Initial research for the following parts is based on thesis work done for the Master's thesis, "Harvesting Plasticity," as part of the University of Tokyo, Advanced Design Studies Laboratory, under professors Yusuke Obuchi and Jun Sato. The "material," "tool," and "positioning and feedback system" portions were developed by Rod (2015), while the "organizational logic", and "structural system" were developed by Clement (2015). These initial studies were significantly developed to produce the case study projects, completed by the University of Tokyo, Advanced Design Studies Laboratory, in concert with Jun Sato Laboratory.

3.2.1 Material

The system is composed of stiff, linear members, connected with thermoplastic strings. The thermoplastic utilized is polylactic acid (PLA), an innocuous substance able to be worked with by humans due its weight, as well as its heating and cooling speed.

In lieu of PLA, initial experiments were conducted that looked at the viability of using a low-grade, starch-based plastic (Rod 2015). Utilizing the research of E.S. Stevens as a starting point, a mixture of vinegar, glycerol, water, and off-the-shelf starch was mixed together and by applying heat, turned into a low grade thermoset plastic (Stevens 2002). It was found that the most effective way of accurately distributing the material was to extrude it through a standard handheld 120 ml syringe with a nozzle of 5 mm diameter (Rod 2015). A CNC machine was also utilized to automate the syringe system and produce various samples. These samples were measured for cracking and shrinkage over several days (Fig. 3.1).

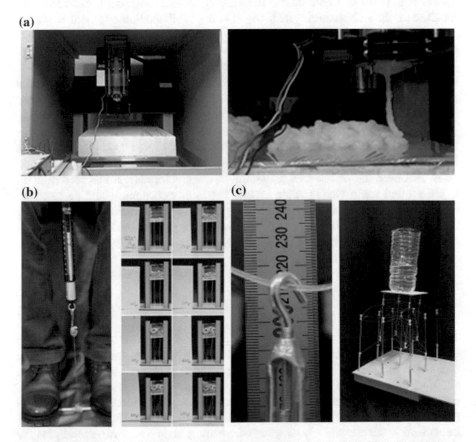

Fig. 3.1 a Syringe in CNC machine (left) and layering plastic extrusion (right) (Rod 2015). **b** Structural tests by Ying Xu, Jun Sato Laboratory. **c** Loading a small-scale model by Kevin Clement, Obuchi Lab

In addition to slow drying times, the material was very brittle and the amount of shrinkage was difficult to effectively control. However, the material exhibited potential in its ease of production, as well as its phase changing behavior. In addition, it was possible to add new material to previously produced plastic through additive layering. As the material dried, it was found that it seemed to possess its own uncontrollable agency.

The concept of utilizing material inconsistencies to produce architecture continued to be considered, even as we began using an industrially made polylactic acid (PLA). In order to improve structural performance and have more control over the material, it was necessary to begin using this higher-grade plastic.

Research by Ying Xu, as part of Jun Sato laboratory, tested the mechanical response of the PLA filament as extruded material, as well as in the specific geometrical configurations of the architectural system itself. Tests were conducted that examined the compressive, tensile, and buckling strength of the PLA, and the Young's modulus was calculated. It was found that the material exhibits characteristics like hard rubber, rather than hard plastic. In addition, when in the specific extruded form required for the architectural system, the filament exhibited limited amounts of stiffness and resistance to buckling and bending. This contrasts with conventional tensile materials used in architectural assemblies, such as steel cables or rope.

One additional factor observed about the material is that due to its thermoplastic nature, as well as the configuration of the system, which utilizes this material as tensioned elements, they are in a state of slow and constant deformation. It was hypothesized that this deformation could be controlled and harnessed to proactively generate form. This hypothesize has been tested heuristically, through the production of the case studies. It has been observed that over time, they do in fact deform in relative accordance to what was hypothesized and shown in computer simulations. However, the precise data pertaining to these deformations has been difficult to collect and study. One possible avenue for further research would be to create more prototypes and then scan and track the motion of the individual members over time. Correlating these changes to predictive digital models could prove fruitful in developing a better understanding of the overall architectural system.

3.2.2 Tool

Introducing the human as a participatory actor in the production of architecture underlined the architectural concept. For this reason, it was necessary to develop an ergonomic tool that could enable the human hand to create full-scale structures. In addition, the production of the system itself, coupled with the weakness and constant deformations of the material, made the introduction of the human necessary in the production process. While robots are good at applying materials in precise locations, humans are better at intuitively grasping where things should go. Furthermore, despite recent advances in artificial intelligence, humans have their own agency that remains distinct from machines.

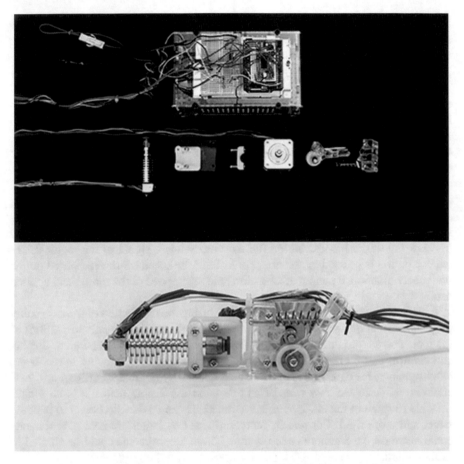

Fig. 3.2 Tool kit of parts (above) and final assembled handheld tool (below) (Photos by Deborah Lopez, Hadin Charbel, Obuchi Lab)

During initial studies with PLA, a commercially available 3D-pen, the 3D-Doodler, was tested. However, its lack of durability, coupled with the inability to control the size of the extrusion, speed of rotator, and heat of the nozzle necessitated the making of a custom 3D-pen designed specifically to deal with these issues (Fig. 3.2).

The tool is composed of readily available and affordable commercial parts and a variety of custom laser cut fitting components. Initially developed as a prototype by Anders Rod in 2015 and used in the creation of "Case Study I," the pen was significantly developed by Ying Xu in 2016 and 2017. The performance and durability of the pen was improved to produce the second case study project.

The pen went through multiple prototyping phases and is still under development. However, the basic design of the pen has been consistent and is as follows: PLA with a 3 mm diameter is inserted through the back of the pen; a clamping mechanism holds this material in place; A stepping motor, connected to a drive shaft and pulley, rotates and pushes the filament through a heat-sink and then a hot-end; The hot-end melts the filament temporarily, allowing the user to draw the material in mid-air before it hardens (Rod 2015).

The size of the extrusion hole, the amount of heat applied, and the speed of extrusion are all variables that need to be calibrated and tailored according to each task to ensure that the process meets both structural and usability criteria. Initial experiments used a standard J-head hole size of 0.5 mm. However, to scale up, structural calculations indicated that an extrusion size between 1.5 and 3.0 mm was desirable.

The hot-end itself consists of a temperature resistor and an inductive heating element. The resistor monitors the current temperature of the hot-end and sends this data to a temperature resistor, whose resistance is read by an Arduino Uno. This value is computed by a script that increases or decreases the amount of current necessary to reach the desired temperature. The Arduino sends a signal to the motor driver when a button is pressed by the user, controlling the speed of the feed based on the previously mentioned factors of extrusion hole size, speed of the motor, and applied heat.

Constant fine-tuning of the pens occurred during the production of the case study projects. Further design studies could be made to enhance the ergonomic qualities of the tool. The addition of extra features to the design, such as a temperature display and control, as well as an ability to change the speed of the drawing in real-time, would enhance the capabilities of the human while they fabricate. Furthermore, tests could be made to see how wide of an extrusion is possible before it is no longer possible to draw in the air. Larger extrusion sizes can enable structures to become larger and more rigid. Conversely, increased size and larger filament requirements could decrease the ergonomics of the tool. These requirements could be studied in further detail.

3.2.3 Organizational Logic

The logic of the material organization was designed to be able to create an open-ended, user-driven system capable of adapting its form to different site or programmatic restraints. The use of the thermoplastic as a connective tissue between stiff compressive members, as well as the need for human involvement, necessitated the use of small elements composed in a certain organization that could account for necessary redundancies in the structure.

The system relies on the stickiness of the PLA and its ability to bind materials together (Clement 2015). As the PLA acts best in tension, a more rigid, linear material was chosen to act in compression. Both timber and acrylic sticks were tested (respectively 3.0 and 2.0 mm in diameter), and as both materials have similar densities, weight, and structural properties, it was decided to proceed utilizing acrylic.

This was due to the idea that it could potentially be substituted with bio-plastic in the future, creating a more holistic approach with regards to materiality.

To control the visual and structural performance of the architecture, the spacing, orientation, location, and angle of each stick was found to be critical. In addition, as the physical production of the architecture involved inaccuracies inherent in human fabrication, structural redundancy, rather than optimization, was utilized to ensure the stability of the construct. This was accomplished through the placement of the sticks utilizing a voxel-grid, wherein each point within the grid acted as the centroid of the stick (Clement 2015). The sticks were placed within this grid and then different patterns of connection were tested both physically and digitally.

As there were many possibilities for differentiation within the system, the production of a digital model became critical to the research process. A digital script was built utilizing Grasshopper, a plug-in for Rhino, and its accompanying Python script module. The script allowed for the creation and exploration of various complex forms at micro and macro scale. It was used to drive the decision-making process with regards to which physical mock-ups were worth testing (Clement 2015).

The design script was built to take a closed BREP as an input, build a voxel-grid within this BREP, and place sticks within this voxel grid. Differentiation was tested both globally, via shaping studies, and locally, by varying the density, angle, and connection pattern of the sticks. Global changes could be affected in two ways. The closed BREP could be deformed via scaling operations, thus creating different spacings of the voxel grid within the shape. Alternatively, Boolean operations could be applied to the BREP itself, resulting in different quantities of points within each row of the voxel grid. These changes were tied to the local parameters of each stick, so that their spacing changed depending on grid deformation, while the change in stick angle was made to be reactive to Boolean subtraction operations (Clement 2015) (Fig. 3.3).

A by-product of this layered approach was that varied patterns and translucencies emerged during the physical production of the architecture. The system was able to absorb inaccuracies of the fabrication process, and adapt to these changes locally, while still maintaining its architectural integrity and aesthetic. The individual actions of each maker were able to be accommodated within the system, resulting in visible differences between areas made by different fabricators.

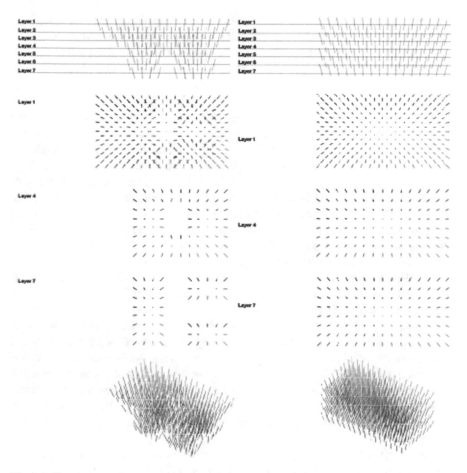

Fig. 3.3 Two organizational logics of compressive members (Clement 2015)

3.2.4 Structural System

Structural research was conducted in partnership with Jun Sato Laboratory, the University of Tokyo with support by Mika Araki for Case Study I, and Ying Xu for Case Study II.

The structural system resembles tensegrity systems developed in the 1960s, wherein a series of stiff compressive members are held together through tensioned cables. However, this structure differs in that the conventional tension cables were substituted with PLA filament. While the filament performs best in tension, its ability to act in compression makes it a semi-rigid, semi-tensegrity system. Conceptually and structurally, tensegrity systems are internally coherent; this system utilizes this principle within a voxel-grid. However, while tensegrity must be precisely calibrated and tensioned to remain stable, the semi-rigidity of the connections, coupled with the layering of modules within a voxel grid, eschews the ideal or the tenable.

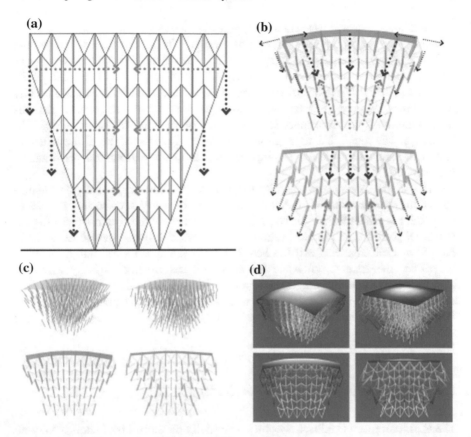

Fig. 3.4 Structural concept and simulations (Clement 2015)

The organizational logic of the system is layered, resulting in a redundant structure, wherein some of the connections are in tension while others buckle. The result is a geometrically non-linear, dynamic structure that adjusts to accommodate local, structural failures that take the form of buckling. These failures are compensated for in other parts of the structure. As new members are added to or subtracted from the system, buckling members can be observed to become taught and once again perform in tension.

A series of physical prototypes and case studies were physically built based on digital models. While it was difficult the ascertain if the structural performance of the smaller prototypes was in accordance with the digital simulations, it was observed that in the case studies, the models appeared to deform in the manner predicted by the models.

Kangaroo and Karamba, plugins for Rhino Grasshopper, were utilized at the beginning of the process. The same digital models were simulated in both programs and the results compared. Although the method of calculation and data output of both programs is different, it was found that the same geometries were predicted to act in the same way (Fig. 3.4).

The digital tests and case studies suggest that both local and global parameters influence the structural performance of the system. The software predicted that the global geometry of the shape would perform best when it tapered outward, which was consistent with our understanding of tensegrity systems in general. The nature of the system is such that the tension members pull the structure inwards, causing it naturally to buckle and fall in on itself. By cantilevering portions of the structure outwards, this inward motion is counterbalanced by the weight of the outward portions (Fig. 3.4). Locally, it was found that angling the sticks away from the center of the structure and positioning them closely together seemed to increase structural performance (Clement 2015).

In addition to parameters regarding the spacing, position, angle, and global geometry of the structure, it was predicted that the size of the PLA extrusion, as well as the pattern in which the acrylic sticks were connected via these extrusions, effected structural performance. Hogan, a structural calculation software developed by a Jun Sato Structural Engineers and Jun Sato Laboratory, showed a correlation between the patterns of stick connection and their structural performance. Digitally, they were able to perform post-buckling analysis on both the freestanding and hanging case studies, after which a nonlinear process analysis was applied. To produce Case Study II, the patterns of connection were altered based on the analysis done with the Hogan simulations.

3.2.5   Positioning and Feedback System

The Positioning and Feedback System was initially developed by Anders Rod, with support from Kosuke Nagata (Rod 2015). It was further developed for the Centre Pompidou case study project, by Jiang Lai, Hirokazu Tei, Alric Lee, Veronika Smetanina, and Ao Yang.

As the human agent is an active participant in the project, a method was developed for multiple fabricators to interact with the digital in real-time. This process can be described as an assembly of "mistakes" that nonetheless can build a coherent, architectural system and form. The form itself is affected by the accumulation of mistakes, so that the final geometry is changed based on the impulses of human actors. As the structure can be added to and subtracted from, the digital feedback system can theoretically continue without end. This adaptable, user-driven architecture seeks to enhance the creative power of anyone that can pick-up the pen and learn the basic processes involved in making the system.

The first iteration of the digital feedback system utilized two off-the-shelf HD webcams to track the motion of the sticks. A camera lens was fitted with an infrared filter. Simultaneously, infrared LEDs were attached to a "tracking clip" that could hold each stick (Rod 2015). The LED light was picked up by the camera lens and located in real-time by utilizing Processing's Open CV library (Fry 2001; Cosout 2008). The initial calibration of the system was done with a genetic algorithm (Yu et al. 2005), which allowed the system to track the position of each member (Rod 2015). In addition, he paired this location tracking system with an accelerometer,

Fig. 3.5 Positioning system and drawing in use (above). Clip with two LEDs lit indicating incorrect stick angle (bottom left), and four LEDs lit indicating correct angle (Rod 2015)

enabling the tracking of the stick angle. Both the location and tracking data was piped through an Arduino micro-processor and sent to Rhino, thus enabling the position and angle of the stick to be compared with the model location in real-time. Blinking lights indicated to the user when they had the location of the stick in the correct place and angle. As the user approached the correct positioning of the stick, the lights would flicker more quickly, and finally become solid once the stick was within an acceptable range of error, which could be adjusted within the script (Fig. 3.5).

The final locations of each stick were calculated and compared with the "ideal" digital locations. It was found that the final locations of each stick were off between 35.2 and 8.8 mm (Rod 2015). However, the overall physical model was structurally stable and appeared the match the digital model in terms of basic stick location, angle, and aesthetic.

Based on the working success of this first model, a more advanced version of the system was researched and tested on a portion of the case study for the Centre

Pompidou. While the initial system was able to register the position and angle of the stick to a certain level of accuracy, human error was high, because fabricators were required to hold the stick at both the correct angle and location simultaneously, and then draw the required connections. In addition, the feedback system was manual and time-consuming, so that only certain sticks around the perimeter of the mock-ups and freestanding case-study were tracked and compared to the original design intent. This meant that the structures were not able to digitally adapt to the physical realities of production. When advancing to the scale of the hanging case study, it was hypothesized that the accumulation of human errors could cause the final structure to fail. For this reason, the tracking system was improved upon so that it would be easier for humans to focus on only placing the stick at the correct location, while automating the position of the stick angle. Lessons learnt from previous built pavilions at the University of Tokyo Advanced Design Studies Laboratory were

Fig. 3.6 Second iteration of positioning system with angle-correcting handheld robot arm (Photos by Deborah Lopez, Hadin Charbel, Obuchi Lab)

feedback and updating systems were successfully integrated into the fabrication process were were used as conceptual touchstones in the development of the project.

The premise of the revised system was as follows:

- The target geometry, composed of compression sticks and tensile strings, is exported. The centroid of the rods is converted into numerical data and sent to a motion-tracking system.
- The worker locates themselves in the physical fabrication space, and their location is detected by a motion-tracking system. This system locates the position of the physical stick and compares it to the digital model, providing real-time feedback of its target location by using an intuitive blinking light code that is integrated into an ergonomic stick holder.
- Once the angular and spatial position of the stick is indicated as being correct, the human holds the stick holder in place and draws the connections to the other sticks using the 3D-pen.
- The clip also doubles as a location checking device. Once the stick has been connected to its neighbors and is in final suspension, a button is pressed, which sends the actual position of the stick back to the digital model and compares the target and actual location of each component.
- After one layer of sticks is placed, the scanned model is analyzed structurally using the Hogan Software developed by Jun Sato Lab. The subsequent layer of sticks is updated based on an algorithm that adopts a specific updating logic that can integrate the error involved in the placement of each component.
- With the errors now digitally compensated for, the fabrication continues, and more "mistakes" are inevitably introduced again. This process repeats for subsequent layers and can in theory be an unending process.

The revised system improved upon the ergonomic interface between the digital and the physical, through the introduction of a handheld, 3-axis mini-robotic arm, tracked by Kinect, that could automatically adjust the angle of each stick in real-time, even as the position and angle of the user's hand changed (Fig. 3.6). This mini robotic arm utilized an accelerometer that could detect the correct angle for the stick and hold it in place while the PLA filament could be drawn. As the angle of each stick was controlled through the robotic arm, it was possible for the human to focus purely on locating the component in the correct position. This made the task of drawing easier and more heuristic. The interaction between material, human, and machine was made explicit through this process.

For positioning, blinking lights were integrated into the handle of the robot-arm, enabling communication with the user to alert them when they had the stick in the correct location. Two Kinects were used to track an IR sticker, placed at the tip of the robotic arm, as close as possible to the centroid of the stick. After the stick was placed, its position was recorded in real time, through the push of a button. Pushing this button sent the actual position of the stick back to the digital model, allowing it to ascertain the drawn, physical position of each stick.

Similar to other pavilion projects built at T_ADS, such as TOCA, human "mistakes" were not perceived as negatives during the construction process. Rather, they

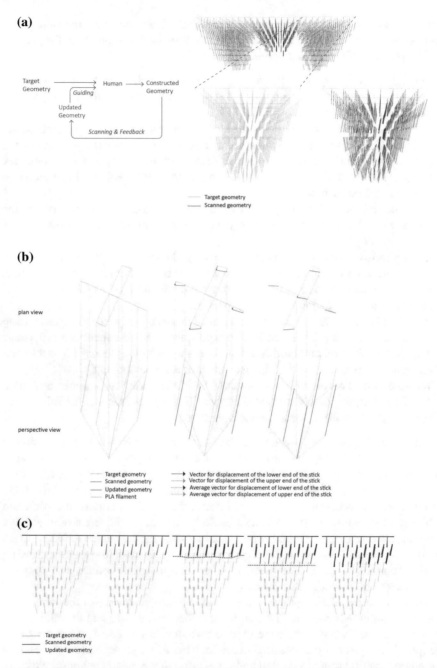

Fig. 3.7 **a** Diagrammatic image, showing how human "mistakes" can be absorbed by scanning the constructed geometry and updating the next geometry to be built. This system was used to produce a small portion of Case Study II. **b** Average vectors for displacement of the scanned geometry from the target geometry are taken to generate the position of the next layer. **c** Conceptual diagram, showing how each layer can be leveled to maintain targeted geometry (Image by Jiang Lai, Obuchi Lab)

were absorbed into an adaptive system. This process involved feeding the tracked stick locations back into the digital model and then updating this model based on the reality of the construction. This reality would update the target geometry, causing local differences to be integrated into the global geometry (Fig. 3.7a). This system was used to produce a small portion of Case Study II. Although the initial concept was to update the geometry based on structural optimization, it was decided to normalize each layer and then evenly distribute the sticks as much as possible, due to processing, time, and design constraints.

The even distribution was achieved by averaging the vectors for displacement of the scanned geometry to the target geometry (Fig. 3.7b). At the same time, each layer was leveled to maintain the overall geometry (Fig. 3.7c). The Guiding-Scanning-Updating loop is intended to repeat and be seamless. However, due to the constraints imposed by hardware capacity and the exigencies of the construction site and schedule, the process was not utilized on the entirety of the case study project. Current research is underway that seeks to use this process throughout the entire fabrication process, thus creating a truly adaptive system that augments human intuition with digital precision, so that non-skilled workers can participate in the process of making architecture.

3.3 Case Studies

Initially, the relationship between the material, tool, organizational logic, structural system and position/feedback system was tested through the fabrication of a series of small scale models. After a table-sized mock-up was produced (Fig. 3.8a), the research speculations were tested in the form of two large-scale prototypes.

3.3.1 Case Study 1—Freestanding Structure

The first case study was a freestanding pavilion, built to stand for 2 weeks, and measuring 2200 mm X 2200 mm X 1600 mm; it was constructed at Ozone Gallery in Shinjuku, Tokyo, Japan (Fig. 3.8b).

As the structure is required to be in tension as it is built, and for ease of drawing, the project was first drawn into place from the top layer and proceeded downwards. The highest layer was top-hung from pre-drilled MDF boards and held in place with small clamps. A timber scaffold held the MDF board in place, and was made to be adjustable, so that the entire structure could be lifted to assist in the drawing process. Due to transportation limits on the size of each scaffold, the pavilion was built as two separate pieces. Upon completion of the pre-fabrication process, the model was transported to an art gallery. It was released as two freestanding constructs, placed onto prefabricated pedestals, and then drawn together on-site utilizing the 3D-pens.

(a)

(b)

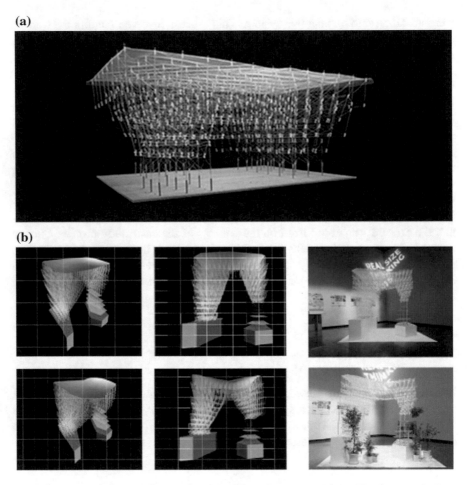

Fig. 3.8 a Free-standing, table-sized mock-up (Photo by Jan Vranovský, Obuchi Lab). **b** Caste study 1, structural simulations and prototype (images/photos by Kevin Clement, Obuchi Lab)

During this process, it was found that the necessity of being able to maneuver one's hand within the space between the sticks to make the connections imposed certain limits on the minimum spacing of the sticks. In addition, when the sticks were placed too far apart, it was difficult to draw a straight, taught line between the members. We found that the minimum spacing between sticks was about 150 mm, while the maximum spacing was about 300 mm. The final pavilion was built with a spacing of 150 mm.

The case study as constructed performed in a similar matter to the structural simulation predicted by Karamba. Due to size and time constraints, there was very little cantilever of the final structure outwards. Karamba predicted that the structure would be weak in its final formal iteration and would deform inwards. This prediction was born out in when we fabricated and joined the final structure together. It was very weak and ultimately fell in on itself, requiring extra support on its perimeter to stand (Fig. 3.8b). Further testing is required to confirm that the use of cantilevering the form outwards can be used to counteract this rotational moment at large scale. The use of the cantilevering was successful in the production of the table-size model (Fig. 3.8a), and the structural simulations suggest that this method can be successful in allowing for large scale iterations of this project to be built that can free stand. Further research and fabrication is required to substantiate the predictions of the simulations and our hypothesis.

3.3.2 Case Study 2—Suspended Structure

The second case study was built to hang in a controlled gallery space for three months and measured 6300 mm X 3200 mm X 2500 mm; it was pre-fabricated at the Ecole Special de Paris and installed at the Centre Pompidou in central Paris, France (Fig. 3.9a).

The pattern and spacing of the sticks was altered to accommodate the different functional requirements of the exhibition. In addition, work by Jun Sato Lab optimized the points required to hang the work from sheets of 5 mm acrylic. To maximize the tensile performance of the PLA in the hanging application, the pattern of connections was essentially flipped. The project was also "drawn-in-place" from the top down, being hung from three sets of timber scaffolds. Once about 4/5 s of the structure was drawn within these scaffolds, it was transported to the art gallery and hung from the ceiling. Workers on scissor lifts were able to learn how to use the pens and then connect the disparate pieces of the structure together, allowing the pieces to perform as one construct (Fig. 3.9b).

For this project, the size of the extrusion nozzle was increased to 2.0 mm, thus causing the PLA, and thus the entire structure, to be much more rigid. However, there was also less springiness in the overall construct. Further research is required to find the optimum balance between extrusion size, stick spacing, and desired springiness.

Fig. 3.9 a Case study 2, suspended structure. **b** (Left) Participatory process, the students taught workers the fabrication method. (Right) Worker participating into the fabrication process (Photos by Deborah Lopez, Hadin Charbel, Obuchi Lab)

3.4 Results and Discussion

The research was able to demonstrate that a synthesized approach to fabrication that integrates material, tool, humans and technology at the outset can result in structures that are locally differentiated but still possess aesthetic coherence. The engagement of humans in the fabrication process created areas in the structures where one could observe differences in the style of the individual makers of the projects (Fig. 3.10).

With regards to material, it was initially speculated that commercialized PLA could be substituted for Poly-L-Lactic Acid, a plastic type similar to PLA, that is made using sugar content from mixed food waste (Sakai 2012). This material exhibits similar physical properties to conventional PLA, and the research would benefit from taking an integrated and holistic approach to the production of the plastic material itself.

Although the case studies were successful in physically influencing space at the human/architectural scale, future experiments would benefit from testing the material and tool limitations. It was observed during tool development that scaling beyond a certain point would require greater amounts of heat applied to a larger, industrial-scale hot end, thus necessitaties a thicker PLA rod. This could potentially the tool ergonomics and cause the production logic to move back towards an automated system.

In relation to the tool specifically, the final pen was able to be used by people from different backgrounds and skill with regards to fabrication. After several minutes of training, they were able to work on making the structure. This opens the drawing process to the general populace. We found that with more practice, people improve their drawing ability, thus making the use of this tool a new type of making enabled by digital technologies. Furthermore, the ease of use and the redundancy of the structure not only allowed for failures to be absorbed into the final output, but inherently integrated drawing instruction through a 'connect the dots' type system. Essentially, each stick had to be connected to all neighboring sticks through a simple set of rules. Through basic procedures and an easy to use portable tool, people can negotiate the spaces between them without being explicitly directed, creating a participatory, networked fabrication space.

Finally, as the structure ultimately has more connections than necessary, there is an embedded possibility to cut away and add more pieces to the structure, forming an open-ended network. However, there are limitations with respect to structural feasibility that must be constantly calibrated and monitored. Future research should seek to study the different types of patterns that emerge through people's behaviors through their bodily/spatial interactions on site and what effects, if any, these different trajectories would have on the final output.

Fig. 3.10 Relationship between material, human agent and tool (above from left to right). Decentralized local workforce (below) (Photos by Deborah Lopez, Hadin Charbel, Obuchi Lab)

3.5 Conclusion

Certain strands of digital fabrication research have explored how material and human inputs can be used projectivity in the production of architecture. Standard industrial methods often seek to create mass produced outputs. Furthermore, both the material and manufacturing processes also need to be consistent. In contrast, it is possible to proactively integrate material inconsistencies into the design process, utilizing these irregularities generatively. The incorporation of humans, not only as labor forces but also as differentiated inputs, into an inclusive digital fabrication process allows for individual expression within systems of production. The introduction of a digital feedback process can modulate the relationship between material agency and individual actors, weaving them into a network that is able to create optimized structures through time. The goal of this project, one still very much in progress, is to create an emancipatory architecture, whose ongoing outputs are the embodiment of collective effort and memories.

Acknowledgements This research was conducted in Obuchi Lab, T-ADS at the University of Tokyo and developed as an extension of the master's thesis 'Harvesting Plasticity' originally conducted by Kevin Clement and Anders Rod. The authors would like to gratefully acknowledge the university and T-ADS for providing a platform for the research; Yusuke Obuchi for his support and guidance; Jun Sato, Mika Araki and Ying Xu for structural analysis; Kengo Kuma and Associates for their support during the production of the final prototype exhibited at the Centre Pompidou; Professors Toshi Kiuchi and Kosuke Nagata for guiding portions of the project; all the students involved in the production throughout the research, including Gilang Arenza, Ratnar Sam, Haruda Uchida, Takahiro Osaka, Hirokazu Tei, Mika Portugaise, Chen Xiaoke, Ruta Stankeviciute, Wu Ziyi, Alric Lee, Nathalia Rotelli, Emi Shiraishi, Veronika Smetanina, Tom Moss, Ao Yang, and Tyler Mcbeth.

References

Bogue M, Dilworth P, Cowen D 3Doodler. http://the3doodler.com/. Accessed 11 Nov 2017

Clement K (2015) Harvesting plasticity section II: digital modeling and simulation of dynamic structural systems and procedural assembly processes for thermoplastic architecture. Unpublished master's thesis, University of Tokyo

Cousot S (2008) OpenCV processing library, École Supérieure d'Art d'Aix-en-Provence

De Paola P (2012) Form follows structure: biomimetic emergent models of architectural production. In: Smith RE, Quale J, Rashia NG (eds) ACSA Conference proceedings, offsite: theory and practice of architectural production

Dillenburger B, Hansmeyer M (2013) The resolution of architecture in the digital age. In: Zhang J, Sun C (eds) Global design and local materialization. CAAD Futures 2013. Communications in computer and information science, vol 369. Springer, Heidelberg

Fry B, Reas C (2001) Processing, MIT Media Lab. https://processing.org/. Accessed June 2015

Kilian, A (2017) Seoul biennale of architecture and urbanism. Autonomous architectural robots—2017 서울도시건축비엔날레 | Seoul Biennale of Architecture and Urbanism 2017. http://seoulbiennale.org/en/exhibitions/thematic-exhibition-nine-commons/sensing/embo died-computation-structure-proposal-of-an-autonomously-acting-structure. Accessed 21 Nov 2017

Kilian, A (2006) Design exploration through bidirectional modeling of constraints

Kudless A (2014) Scripted Movement 1 «MATSYS. MATSYS RSS. 2014. http://matsysdesign.co m/category/projects/scripted-movement-1/. Accessed 21 Nov 2017

Lagerkvist, S, von der Lancken C, Lindgren A, Sävström K (2005) Overview of sketch furniture project. MoMA Multimedia 2005. http://www.moma.org/explore/multimedia/audios/37/856

Lim S, Buswell RA, Le TT, Austin SA, Gibb AGF, Thorpe T (2012) Developments in construction-scale additive manufacturing processes. Autom Constr 21: 262–68. https://doi.org/10.1016/j.aut con.2011.06.010

Lopez D, Charbel H, Obuchi Y, Sato J, Igarashi T, Takami Y, Kiuchi T (2016) Human touch in digital fabrication. In: ACADIA 2016 POSTHUMAN FRONTIERS: data, designers, and cognitive machines, pp 382–393

Malé–Alemany M, Portell J (2017) FABbots: research in additive manufacturing for architecture. In Gramazio F, Kohler M, Langerber S (eds) Fabricate: negotiating design & making. UCL Press, London, pp 207–215

Menges A (2018) Design research agenda. http://www.achimmenges.net/?p=4897. Accessed 28 Feb 2018

Pegna J (1997) Exploratory investigation of solid freeform construction. Autom Constr 5(5):427–37. https://doi.org/10.1016/s0926-5805(96)00166-5

Retsin G, Garcia MJ (2016) Discrete computational methods for robotic additive manufacturing: combinatorial toolpaths. In: ACADIA 2016 POSTHUMAN FRONTIERS: data, designers, and cognitive machines, pp 332–341

Rod AP (2015) Harvesting plasticity section I: development of hand-held plastic extrusion tool and motion tracking and interface system for thermoplastic architecture assembly processes. Unpublished master's thesis, University of Tokyo

Sakai K, Poudel P, Shirai Y (2012) Total recycle system of food waste for Poly-L-Lactic acid output. In: Advanced in applied biotechnology

Stevens, ES (2002) Green plastics an introduction to the new science of biodegradable plastics. University Press, Princeton, pp 83–88

Yoshida H, Igarashi T, Obuchi Y, Takami Y, Sato J, Araki M, Miki M, Nagata K, Sakai K, Igarashi S (2015) Architecture-scale human-assisted additive manufacturing. ACM Trans Graphics (TOG)—Proc ACM SIGGRAPH 34(4):1–8

Yu KY, Wong KH, Chang MMY (2005) Pose estimation for augmented reality applications using genetic algorithm, The Chinese University of Hong Kong

Chapter 4
From Architectured Materials to Large-Scale Additive Manufacturing

Justin Dirrenberger ⓘ

Abstract The classical material-by-design approach has been extensively perfected by materials scientists, while engineers have been optimising structures geometrically for centuries. The purpose of architectured materials is to build bridges across the microscale of materials and the macroscale of engineering structures, to put some geometry in the microstructure. This is a paradigm shift. Materials cannot be considered monolithic anymore. Any set of materials functions, even antagonistic ones, can be envisaged in the future. In this paper, we intend to demonstrate the pertinence of computation for developing architectured materials, and the not-so-incidental outcome which led us to developing large-scale additive manufacturing for architectural applications.

4.1 Introduction

Materials are ubiquitous in nature and man-made applications, from electronics to transportation or energy, but also biomedical, defence, architecture or construction. Materials have such a predominance in human existence that prehistoric periods were named after materials use (Hummel 2004). Technological progress and its impact on everyday's life are intimately linked with our capacity to control matter in the form of industrialised materials. The classical material-by-design approach has been extensively perfected by materials scientists, while civil engineers and architects have been optimising structures geometrically for centuries. The purpose of architectured materials is to build bridges across the microscale of materials and the macroscale of engineering structures, to introduce some geometry within the microstructure. This is a paradigm shift in the sense that materials cannot be considered as monolithic artefacts anymore. Any set of material functions, even antagonistic ones (Steeves et al.

J. Dirrenberger (✉)
Laboratoire PIMM, Arts et Métiers-ParisTech, Cnam, CNRS,
151 boulevard de l'Hôpital, 75013 Paris, France
e-mail: justin.dirrenberger@ensam.eu

J. Dirrenberger
XtreeE, 18/20 rue du Jura, CP 40502, 94623 Rungis Cedex, France

© Springer International Publishing AG, part of Springer Nature 2018
H. Bier (ed.), *Robotic Building*, Springer Series in Adaptive Environments,
https://doi.org/10.1007/978-3-319-70866-9_4

2007), can be envisaged in the future. The development of architectured materials is intrinsically transdisciplinary, on the fringes of physics, chemistry, and mechanical engineering, but also biology, computer science, architecture, design, etc. In this work, we intend to demonstrate the pertinence of a computational approach for developing architectured materials, and the not-so-incidental outcome which led us to developing large-scale additive manufacturing for architectural applications. This paper is an extended version of a preliminary conference abstract submitted for the Next Generation Building workshop held at TU Delft in 2016 (Dirrenberger 2017).

The chapter is organised as follows: an introduction to architectured materials is given in Sect. 4.2. An overview of the computational tools relevant for the development of such materials is presented in Sect. 4.3. The translation of architectured materials concepts to the architectural scale corresponds to Sect. 4.4. Finally, conclusions and perspectives are postponed to Sect. 4.5.

4.2 Architectured Materials

Many industrial applications require new materials with enhanced specific properties, i.e. performance per unit of mass; this is especially true for the transportation and biomedical sectors. For instance, the automotive industry, with its ever increasing requirements regarding passenger safety and fuel consumption, is an edifying example: nowadays, classical steel-based material solutions are being challenged by new lightweight aluminium alloys and advanced composites. A response from steel manufacturers was the development of advanced high-strength steels (AHSS) for yield strength designed parts in order to reduce both thickness and mass. AHSS with very high strength (1000 MPa and more) commonly exhibit poor sheet formability hence limiting the mass reduction attainable. This is especially true for martensite-based AHSS such as dual-phase (DP), complex phase (CP) and martensitic (MS) steels, depending on their martensite content. The higher the martensite content, the lower the thin-sheet formability. In order to mitigate failure and tearing of thin-sheets during deformation-based forming (stamping, blanking, etc.), a possible solution is to rely on the concept of localised heat treatment in order to soften the material where needed by locally annealing martensite, hence changing the local yield strength/ductility trade-off. Introducing such geometrical discontinuities in terms of material behaviour is characteristic of architectured materials.

Architectured materials are a rising class of materials that bring new possibilities in terms of functional properties, filling the gaps and pushing the limits of Ashby's materials performance maps (Ashby and Bréchet 2003). The term architectured materials encompasses any microstructure designed in a thoughtful fashion, such that some of its materials properties, e.g. yield strength/density, have been improved in comparison to those of its constituents, due to both structure and composite effects, which depend on the multiphase morphology, i.e. the relative topological arrangement between each phase (Ashby and Bréchet 2003; Bouaziz et al. 2008; Bréchet and Embury 2013). Localised material processing methods, such as additive manufacturing, or localised laser heat treatment in the case of AHSS, appear as natural candidates for developing architectured materials.

There are many examples: particulate and fibrous composites, foams, sandwich structures, woven materials, lattice structures, etc. with different objectives. For instance, developing architectured porous materials for structural, acoustic and insulation properties (Caty et al. 2008; Fallet et al. 2008), entangled monofilament of perlitic steel (Courtois et al. 2012), sandwich composite structures (Kolopp et al. 2013), segmented interlocking structures (Djumas et al. 2016), woven and non-woven textile composites (Mezeix et al. 2009; Lewandowski et al. 2012; Dirrenberger et al. 2014), crumpled metallic foils (Bouaziz et al. 2013), etc. Architectured materials have been an exciting research topic in recent years, especially regarding the processing of architectured metallic foams (Brothers and Dunand 2006; Erk et al. 2008), multi-scale architectured ceramics (Mirkhalaf et al. 2014), or architectured metallic sheets (Embury and Bouaziz 2010; Chéhab et al. 2009). Apart from processing, most efforts have been focused on the modelling of architectured materials. In particular, the pioneering works done on truss-structures (Deshpande et al. 2001) and metallic foams (Deshpande and Fleck 2000) led to useful results for the modelling of architectured materials (Fleck et al. 2010). The mechanical modelling of bio-inspired architectured materials has been pursued successfully, yielding results for multi-layered materials that could be used for modelling architectured metallic sheets (Turcaud et al. 2011; Stoychev et al. 2012). Other works demonstrated the interest for designing recursive or nested material architectures to achieve enhanced specific mechanical properties (Ajdari et al. 2012; Rayneau-Kirkhope et al. 2012); finally, elastic instabilities have been considered for shape-generation of architectured materials (Bertoldi et al. 2010; Shim et al. 2013).

One can play on many parameters in order to obtain architectured materials, but all of them are related either to the microstructure or the geometry. Parameters related to the microstructure can be optimised for specific needs using a materials-by-design approach, which has been thoroughly developed by chemists, materials scientists and metallurgists. Properties improvements related to microstructural design are intrinsically linked to the synthesis and processing of materials and are therefore due to micro and nanoscale phenomena, taking place at a scale ranging from 1 nm to 10 μm. This scale is below the scope of the present chapter, in terms of topology optimisation, but has been extensively studied in the literature (Embury and Bouaziz 2010; Olson 2001; Freeman 2012).

Until now, most architectured materials in the literature have been obtained empirically. By capitalising on the concept of localised processing of thin structures, our efforts at the PIMM laboratory[1] have been focused on developing systematic tools to determine materials architecturation patterns, for a given set of requirements. For instance, these patterns come as an output from a computational shape-optimisation loop developed around a heuristic generative geometry module, based on cellular automata (CA), and a cost function evaluation module, based on the finite element (FE) method. This cost function has to be minimised for given constraints (Bendsøe

---

[1] http://pimm.paris.ensam.fr/en.

and Sigmund 2004). In the case of linear elastic problems, the cost function can be related to the elastic energy density computed by numerical homogenisation at the scale of the structure (Allaire 2002). The cost function value is then used as a feedback, and an optimised shape is generated accordingly following a specific evolution algorithm. The spatial resolution associated with the shape-optimisation can be chosen to be relevant for localised laser treatment, i.e. 1 mm, corresponding to a representative scale of the underlying microstructure (Bironeau et al. 2016); this allows us to use homogenised behaviour for each grid-cell in the simulation (Dirrenberger 2012).

From a macroscopic viewpoint, parameters related to the geometry have mainly been the responsibility of structural and civil engineers for centuries: to efficiently distribute materials within structures. An obvious example would be the many different strategies available for building bridges. At the millimetre scale, materials can be considered as structures, i.e. one can enhance the bending stiffness of a component by modifying its geometry while keeping the lineic mass (for beams) or surfacic mass (for plates) unchanged (Weaver and Ashby 1996). On the other hand, one might need a lower flexural strength for specific applications, with the same lineic and/or surfacic masses. This can be achieved with strand structures, i.e. by creating topological interfaces in the material. Processing remains the key technological issue for further development of architectured materials, and progress is made every day in this direction at the lab scale, as it was done in (Schaedler et al. 2011) by using a sequence of several processing techniques in order to fabricate ultralight metallic microlattice materials (Schaedler and Carter 2016). There is still a long way to go for the industry to actually apply architectured materials in product manufacturing.

Architectured materials lie between the microscale and the macroscale. This class of materials involves geometrically engineered distributions of microstructural phases at a scale comparable to the scale of the component (Ashby and Bréchet 2003; Bouaziz et al. 2008; Bréchet and Embury 2013), thus calling for enriched models of continuum mechanics in order to determine the effective properties of materials (Geers and Yvonnet 2016; Matouš et al. 2017), e.g. generalised continua theories, in order to describe the behaviour of architectured materials, such as strain-gradient elasticity (Auffray et al. 2015), and strain-gradient plasticity. This topic has been especially fruitful these last few years in the mechanics of materials community (Lebée and Sab 2012; Trinh et al. 2012; Chen et al. 2014; Auffray et al. 2015; Placidi and El Dhaba 2015; Rosi and Auffray 2016; Andreaus et al. 2016; Placidi et al. 2017; dell'Isola et al. 2017); this results in the availability of versatile models able to describe the various situations encountered with architectured materials. Given mature processing techniques, architectured materials are promised to a bright future in industrial applications due to their enticing customisable specific properties and the opportunity of multifunctionality.

When considering actual applications, one engineering challenge is to predict the effective properties of such materials; computational homogenisation using finite element analysis is a powerful tool to do so. Homogenised behaviour of architectured materials can thus be used in large structural computations, hence enabling the dissemination of architectured materials in the industry. Furthermore, computational homogenisation is the basis for computational topology optimisation (Allaire 2002; Bendsøe and Sigmund 2004; Guest and Prévost 2006; Challis et al. 2008; Xu et al. 2016; Vicente et al. 2016; Xu et al. 2016; Salonitis et al. 2017; Asadpoure et al. 2017; Khakalo and Niiranen 2017; Wang et al. 2017) which will give rise to the next generation of architectured materials as it can already be seen in the works of (Laszczyk et al. 2009; Andreassen et al. 2014; Körner and Liebold-Ribeiro 2015; Hopkins et al. 2016; Kotani and Ikeda 2016; Ghaedizadeh et al. 2016; Ren et al. 2016; Liu et al. 2016; Dalaq et al. 2016).

Materials science comes from the following fact: microstructural heterogeneities play a critical role in the macroscopic behaviour of a material (Besson et al. 2010; Bornert et al. 2001; Jeulin and Ostoja-Starzewski 2001; François et al. 2012; Torquato 2001; Ostoja-Starzewski 2008). Constitutive modelling, thanks to an interaction between experiments and simulation, is usually able to describe the response of most materials in use. Such phenomenological models, including little to no information about the microstructure, cannot necessarily account for local fluctuation of properties. In that case, the material is considered as a homogeneous medium. Studying the behaviour of heterogeneous materials involves developing enriched models including morphological information about the microstructure (Smith and Torquato 1988; Yeong and Torquato 1998; Torquato et al. 1998; Decker et al. 1998; Jeulin 2000; Kanit et al. 2006; Peyrega et al. 2011; Jean et al. 2011; Escoda et al. 2015). These models should be robust enough to predict effective properties depending on statistical data (volume fraction, n-point correlation function, etc.) and the physical nature of each phase or constituent. As a matter of fact, advanced models are often restricted to a limited variety of materials. Although isotropic and anisotropic polycrystalline metals, for instance, have been extensively studied by the means of both analytical and computational tools (Cailletaud et al. 2003; Kanit et al. 2003; Madi et al. 2007; Berdin et al. 2013; Fritzen et al. 2013; Hor et al. 2014; Kowalski et al. 2016; Amodeo et al. 2016; Schindler et al. 2017), architectured materials bring up new challenges regarding the determination of effective properties.

The development of architectured materials is related to the availability of appropriate computational tools for both design and modelling, but also for computerised manufacturing as for the various additive manufacturing techniques considered to produce architectured materials.

4.3   Computation for Design, Modelling, and Manufacturing

The merit of additive manufacturing is often summarised as its ability to produce shapes that result from a topology optimisation process. Topology optimisation aims at attaining the most efficient structure for a given set of requirements. It is a long-standing topic of research and development for engineers that can be traced back to the seminal work of (Michell 1904) on frame structures one century ago, or even 30 years earlier with (Lévy 1874), who gave the first proof for determinacy in statics for single-load trusses. The topic of optimisation has been active ever since. Nowadays, optimality in terms of industrial design is becoming more and more critical due to scarcity of material resources and the need for lightweight structures.

This technique has become well-established in the field of structural mechanics, especially when associated with FE simulation. Classical methods (Bendsøe and Kikuchi 1988; Rozvany 1995; Duysinx and Bendsøe 1998), such as SIMP (Bendsøe and Sigmund 2004) (Solid Isotropic Material with Penalisation) rely on node-based values to evaluate and optimise the geometry, i.e. the number of design variables is equal to the number of elements available in the model at initialisation. Then, the optimisation procedure consists in determining at each element if it should either stay a material element or become a void element, i.e. be removed. This technique has been applied to different scales: for instance with regards to the design of efficient building structures (Cui et al. 2003), or as a tool for designing micro- and nano-architectured materials (Zhou and Li 2008).

Most computational approaches for topology optimisation used in engineering are gradient-based, they are also known as local approaches. In recent years, so-called global approaches emerged, and are currently subject to epistemological controversy in the optimisation community due to the lack of proof for global convergence, as well as inefficiency in comparison with classical gradient methods (Sigmund 2011; Le Riche and Haftka 2012). Although the fact that being based on heuristics has been held against global approaches in structural mechanics, heuristics itself should not be considered a shortcoming but rather an epistemological hypothesis.

The computational framework being developed at PIMM rely on a global topology-optimisation approach making use of cellular automata, as well as FE for evaluating of the cost function, which has to be minimised for given constraints (Bendsøe and Sigmund 2004). This evaluation step is straightforward since it consists in performing a FE computation on the generated model using predefined local constitutive behaviours, and averaging the response of the structure, to a given set of boundary conditions and applied loads, by the computational homogenisation method (Dirrenberger et al. 2011, 2012, 2013, 2014). The topology optimisation step is somewhat more difficult as choices have to be made with regards to the many approaches available and the type of problem being dealt with. Several reviews are available on this topic, see for instance (Bendsøe and Sigmund 2004; Rozvany 2009; Eschenauer and Olhoff 2001). Most developments in topology optimisation dealt with the efficiency of structures, i.e. minimising the mass of materials while optimising the elastic stiffness of a structure under a given load, which means choosing between void and matter for any given point in the design space, either continuous (Allaire 2002) or discrete (Bendsøe and Sigmund 2004).

Architecturation patterns can be either continuous, discrete, periodic or random, therefore cellular automata (CA) seem like natural candidates for generating them. CA correspond to an evolving structure based on a regular lattice, they are characterised by 5 properties: lattice geometry and dimensionality, cellular neighbourhood, cell state, local rule of transition, and boundary conditions. CA have already been used for topological optimisation, but not for solving pattern-type problems (Hajela and Kim 2001; Missoum et al. 2005; Hopman and Leamy 2010). An alternative method, the hybrid cellular automata (HCA) method, was proposed by (Tovar et al. 2004) by taking the best of both CA and SIMP. This method has been developed thoroughly for structural optimisation (Tovar et al. 2007). A critical improvement to the CA approach was the implementation of multigrid methods in order to accelerate the convergence during optimisation (Kim et al. 2004; Zakhama et al. 2009). The development of genetic algorithms in structural optimisation is known as evolutionary structural optimisation (ESO). It was first proposed by (Xie and Steven 1993; Chu et al. 1996) in the early 1990s, but suffered from various drawbacks (lack of convergence, mesh-dependency) which were partially overcome by the bi-directional evolutionary structural optimisation (BESO) approach developed by the same team a decade later (Huang and Xie 2007). Other developments have been undertaken regarding genetic algorithms, e.g. in terms of multi-criterion optimisation (Canyurt and Hajela 2010).

Therefore, a framework based on CA with a local rule of transition using the BESO approach, along with multigrid implementation appears like an appropriate option to design architectured materials, especially by the ability of such framework to deal with multiple scales of topology optimisation. In order to be fully efficient, the multiscale optimisation scheme must comply with optimising multiple anisotropic materials with nonlinear elastoplastic behaviour, which will yield nonlinearities in structural response of the architectured materials for various sets of requirements/cases of application, in terms of ductile fracture or fatigue properties (Torabian et al. 2016a, b, 2017a, b).

Generating and modelling shapes for additive manufacturing follows specific rules, coming from both processing constraints, e.g. layer thickness, product dimensions, etc., and the functional properties of the produced part, e.g. mechanical strength, thermal conductivity, etc. An usual and straightforward method for generating an additive manufacturing building path is to use a 3D-to-2D slicing software. It consists in slicing the 3D shape, i.e. computer-aided design (CAD) file, of an object into flat thin layers of constant thickness which can be layered one up onto the other, i.e. computer-aided manufacturing (CAM) file. This results in a cantilever-method strategy. Each layer is then made of a contour line, as well as a filling pattern such as a honeycomb structure or a space-filling curve (Peano curve, Hilbert curve, etc.); the filling density can be adjusted for given requirements. This method is well-established for small-scale additive manufacturing.

4.4   Development of the Large-Scale Additive Manufacturing

The industrial applications of small-scale additive manufacturing in production (in opposition to prototyping) concern either geometrically complex products in high value-adding sectors (biomedical, sports, aerospace) and/or parts made of costly materials, typically alloys which include Titanium, Nickel and/or Chrome in their composition. This type of additive manufacturing is usually powder-based, which is intrinsically costly, but enables very good spatial accuracy and microstructural control. Technologies such as direct metal laser sintering (DMLS), selective laser sintering (SLS), electron beam melting (EBM) or direct metal deposition (DMD) are taking off in high-end industries due to its cost-saving possibilities both in terms of reducing the number manufacturing steps and the amount of materials used (Gutowski et al. 2009; Allwood et al. 2011; Yoon et al. 2014; Baumers et al. 2016; Ford and Despeisse 2016).

The DEMOCRITE project (2014–2015, funded by HESAM Université) aimed at transposing such possibilities within an architectural context, i.e. developing large-scale additive manufacturing with the same level of material performance than conventional processing methods and comparable accuracy as DMLS, i.e. a spatial resolution circa 0.1% of the building size, about 1 mm for a metre-wide printer. Our group at the PIMM lab explored the potential applications of large-scale additive manufacturing techniques to civil engineering structures. Thanks to this large-scale processing technique, one could apply the concept of architectured materials within an architectural context; multifunctional properties can be achieved for structural elements, by optimising geometry and material composition, as it was done in Gosselin et al. (2016), for an enhanced thermal insulation case study regarding a structural wall element produced by large-scale additive manufacturing. An example of topology optimisation based on the SIMP approach, performed during the DEMOCRITE project is given in Fig. 4.1. The initial optimisation results (Fig. 4.1a) were adapted and refined in order to fit the additive manufacturing constraints as well as the architectural scenario. The rendering shown on Fig. 4.1b actually includes the visible layers of printed concrete resulting from the building process.

Based upon an understanding of the limitations identified in previous projects present in the literature, the DEMOCRITE project dealt with the large-scale additive

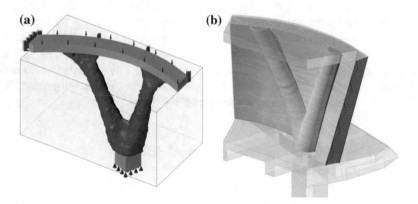

Fig. 4.1 An example of implementation of topology optimisation in the DEMOCRITE project: **a** Boundary conditions and results for the structural optimisation problem; **b** Rendering of the actual multifunctional wall including both the structural and thermal insulation parts

manufacturing of selective deposition for ultra-high performance concrete (UHPC) (Duballet et al. 2015). The 3D printing process involved is based on a Fused Deposition Modelling-like technique, in the sense that a material is deposited layer-by-layer through an extrusion printhead. The project also explored the possibilities offered by CAD and optimisation, and their integration within the product design process in the case of large-scale additive manufacturing. Thus, the introduced technology succeeded in solving many of the problems that could be found in the literature. Most notably, the process enabled the production of 3D large-scale complex geometries, without the use of temporary supports, as opposed to 2.5D examples found in the literature for concrete 3D printing (Khoshnevis 2004; Buswell et al. 2007; Cesaretti et al. 2014).

According to Gosselin et al. (2016), the concept of freeform commonly used in the literature is not adequate nor sufficient for describing concrete 3D printing. For a given printing process and automation complexity, one can attain specific types of topologies within a given time-frame and performance criterion for the material and/or structure to be built. Design conditions for large-scale additive manufacturing depend on many other parameters than just the properties of extruded cementitious materials; parameters such as the printing spatial resolution, overall size of parts to be printed, the environment, the presence of assembling steps, etc. A tentative classification of such relationships between geometrical complexity, processing, and design is proposed in Duballet et al. (2017). The cantilever strategy, which is characteristic of small-scale additive manufacturing, is not appropriate for large-scale printing since it does not take into account the processing constraints and their impact on the performance of the printed structure. The building path should be adapted and optimized based on simulation results in order to take into account constraints and to exhibit more robustness for complex geometries.

The processing constraints depend mostly on the fresh material properties in its viscous state, as well as early-age behaviour, in interaction with the building strategy and the stiffness of the structure being built. On the other hand, functional requirements will depend on the properties of the hardened material, its durability (Lecampion et al. 2011), as well as the structural geometry for effective stiffness, and other functional properties such as thermal and sound insulation. See Gosselin et al. (2016) for a geometrically induced thermal insulation case study. Both types of constraints have to be considered at the design stage.

Printing path generation is a critical step during the design phase. There are two main approaches to tool-path generation in the context of 3D printing: (1) 3D-to-2D slicing, which is by far the most common method adopted, yields planar layers of equal thickness built on top of each other. This approach is not optimal from a design and structural viewpoint as it will induce cantilevers when two consecutive layers have different sizes and limit the attainable geometries; (2) the tangential continuity method, which has been introduced in Gosselin et al. (2016) in order to optimise the structure being built by creating layers of varying thickness. These layers exhibit a maximised surface area of contact between each other, hence stabilising the overall structure. Moreover, this method is actually exploiting the possibilities

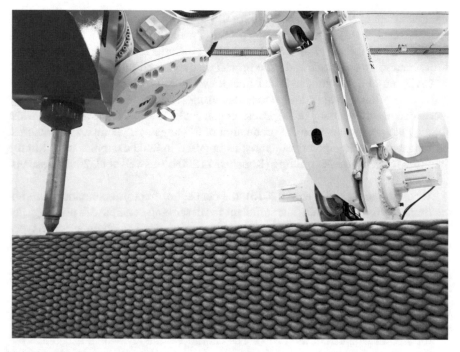

Fig. 4.2 The stand-alone concrete additive manufacturing process developed during the DEM-OCRITE project, currently being implemented commercially by XtreeE

of the process in terms of printing speed and flow for generating variations in the layer thickness. Capitalising on the relative success of the DEMOCRITE project, a spin-off company, XtreeE,[2] was created in order to develop and commercialise the 3D printing technology introduced. The large-scale robotic additive manufacturing process for concrete presented in Fig. 4.2 was developed during the DEMOCRITE project before being applied within the construction industry by XtreeE.

Figure 4.3 shows a 3D-printed 4 metre-high post supporting the playground roof of a school in Aix-en-Provence, France, produced by XtreeE. The structure was made of two parts: the 3D-printed envelope, used as a lost formwork, and the core, made of conventional cast concrete. The total printing time was 15 h and 30 min.

A driving force for additive manufacturing is its ability to produce more complex 3D shapes in comparison to casting or subtractive processes. This complexity allows to design optimal structures based on topology optimisation techniques. One of the main current challenges is to modify optimisation algorithms in order to account for the additive manufacturing constraints, especially with regards to the processing parameters and structural stability while printing. A possible answer to these challenges would be to consider the multiphysics phenomenon aspect of 3D printing, which involves the elastic stability of the overall structure being built, the kinetics

---

[2]http://www.xtreee.com.

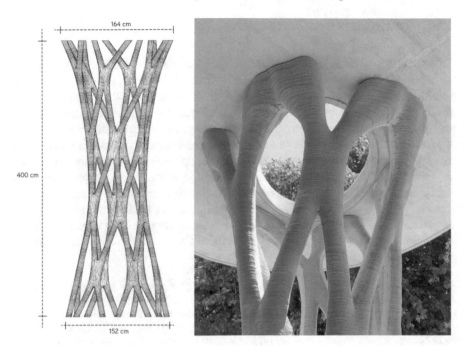

Fig. 4.3 Close-up view of a 4 metre-high post in Aix-en-Provence, France. *Source* www.xtreee. com

of hydration, the evolving viscoplasticity of fresh cement, the evolution of temperature within the printing environment, etc. As a matter of fact, all these physical problems, with multiple time and space scales, can be modelled on their own, but coupling them generates complexity and uncertainty regarding the process of 3D printing. Therefore, efforts should be concentrated on understanding and modelling the printing process in its multiple physical aspects, only then optimisation will be fully integrated with the processing, which would virtually change the way 3D printed structures are conceived today. An attempt at multi-objective optimisation for thermal, mechanical and economical criteria, in the context of a hybrid large-scale concrete 3D printing and assembling process, was proposed by Duballet et al. (2018). It consists in fabricating insulating blocks, which would be assembled by robots in order to form a layer of printing support for the mortar to be extruded; at the end, tie columns and ring beams will ensure the masonry confinement. This approach is rather new in comparison with most examples of large-scale concrete 3D printing available (Labonnote et al. 2016; Duballet et al. 2017). Indeed, the future of large-scale additive manufacturing in architecture and construction might reside in smarter, more parsimonious use of 3D printing on specific parts of the printed object, where it is the most pertinent in order to take advantage of both the material properties and morphology of the structure being built.

4.5 Conclusions and Perspectives

An overview was given regarding the concept of architectured materials in general, i.e. highlighting the importance of geometry at multiple scales, as well as the computational tools associated with their development. In fact, this development is made possible by the computerisation of both the designing step, and robotisation of the manufacturing step. This automation of the process theoretically allows for a higher accuracy and infinite customisation of materials and structures according to a given set of requirements. In this context, the top-down approach of architectured materials, i.e. going from the application down to the design of materials, becomes generalisable, hence yielding new opportunities for the built environment sector, e.g. in architectural design, structural engineers' calculations, or the logistics of a construction site, if building information modelling (BIM) is integrated. From an application viewpoint, one of the most critical point is indeed to be able to incorporate robotic actions within the BIM model.

Acknowledgements This work is part of the DEMOCRITE (Large-scale additive manufacturing platform) Project PNM-14-SYNG-0002-01, as well as the SCOLASTIC (Systematic Computational Optimisation and Local Laser Processing for Steel-Based Architectured Materials) Project ANR 16-CE08-0009. The author would like to gratefully acknowledge ANR (Agence Nationale de la Recherche), and *heSam Université* for financial support through its *Paris Nouveaux Mondes* program.

References

Ajdari A, Jahromi BH, Papadopoulos J, Nayeb-Hashemi H, Vaziri A (2012) Hierarchical honeycombs with tailorable properties. Int J Solids Struct 49(11–12):1413–1419
Allaire G (2002) Shape optimization by the homogenization method. Springer
Allwood JM, Ashby MF, Gutowski TG, Worrell E (2011) Material efficiency: a white paper. Resour Conserv Recycl 55(3):362–381
Amodeo J, Dancette S, Delannay L (2016) Atomistically-informed crystal plasticity in MgO polycrystals under pressure. Int J Plast 82:177–191
Andreassen E, Lazarov B, Sigmund O (2014) Design of manufacturable 3D extremal elastic microstructure. Mech Mater 69(1):1–10
Andreaus U, dell'Isola F, Giorgio I, Placidi L, Lekszycki T, Rizzi NL (2016) Numerical simulations of classical problems in two-dimensional (non) linear second gradient elasticity. Int J Eng Sci 108:34–50
Asadpoure A, Tootkaboni M, Valdevit L (2017) Topology optimization of multiphase architected materials for energy dissipation. Comput Methods Appl Mech Eng 325:314–329
Ashby MF, Bréchet Y (2003) Designing hybrid materials. Acta Mater 51:5801–5821
Auffray N, Dirrenberger J, Rosi G (2015) A complete description of bi-dimensional anisotropic strain-gradient elasticity. Int J Solids Struct 69–70:195–210
Baumers M, Dickens P, Tuck C, Hague R (2016) The cost of additive manufacturing: machine productivity, economies of scale and technology-push. Technological Forecasting & Social Change
Bendsøe M, Kikuchi N (1988) Generating optimal topologies in structural design using a homogenization method. Comput Methods Appl Mech Eng
Bendsøe M, Sigmund O (2004) Topology optimization. Springer

Berdin C, Yao ZY, Pascal S (2013) Internal stresses in polycrystalline zirconia: microstructure effects. Comput Mater Sci 70:140–144

Bertoldi K, Reis P, Willshaw S, Mullin T (2010) Negative Poisson's ratio behavior induced by an elastic instability. Adv Mater 22(3):361–366

Besson J, Cailletaud G, Chaboche J-L, Forest S, Blétry M (2010) Non-linear mechanics of materials, vol 167. Solid Mechanics and Its Applications. Springer, Heidelberg

Bironeau A, Dirrenberger J, Sollogoub C, Miquelard-Garnier G, Roland S (2016) Evaluation of morphological representative sample sizes for nanolayered polymer blends. J Microsc 264(1):48–58

Bornert M, Bretheau T, Gilormini P (2001) Homogénéisation en mécanique des matériaux, Tome 1: Matériaux aléatoires élastiques et milieux périodiques. Hermès

Bouaziz O, Bréchet Y, Embury JD (2008) Heterogeneous and architectured materials: a possible strategy for design of structural materials. Adv Eng Mater 10(1–2):24–36. https://doi.org/10.1002/adem.200700289

Bouaziz O, Masse JP, Allain S, Orgéas L, Latil P (2013) Compression of crumpled aluminum thin foils and comparison with other cellular materials. Mater Sci Eng A: Structural Materials: Properties, Microstructure and Processing 570:1–7

Bréchet Y, Embury JD (2013) Architectured materials: expanding materials space. Scripta Mater 68(1):1–3

Brothers AH, Dunand DC (2006) Density-graded cellular aluminum. Adv Eng Mater 8(9):805–809

Buswell R, Soar R, Gibb A, Thorpe A (2007) Freeform construction: mega-scale rapid manufacturing for construction. Automation Construction 16:224–231

Cailletaud G, Forest S, Jeulin D, Feyel F, Galliet I, Mounoury V, Quilici S (2003) Some elements of microstructural mechanics. Comput Mater Sci 27:351–374

Canyurt OE, Hajela P (2010) Cellular genetic algorithm technique for the multicriterion design optimization. Struct Multidiscip Optim 40:201–214

Caty O, Maire E, Bouchet R (2008) Fatigue of metal hollow spheres structures. Adv Eng Mater 10(3):179–184

Cesaretti G, Dini E, Kestelier XD, Colla V, Pambaguian L (2014) Building components for an outpost on the Lunar soil by means of a novel 3D printing technology. Acta Astronaut 93:430–450

Challis VJ, Roberts AP, Wilkins AH (2008) Design of three dimensional isotropic microstructures for maximized stiffness and conductivity. Int J Solids Struct 45:4130–4146

Chéhab B, Zurob H, Embury D, Bouaziz O, Bréchet Y (2009) Compositionally graded steels: a strategy for materials development. Adv Eng Mater 11(12):992–999

Chen Y, Liu XN, Hu GK, Sun QP, Zheng QS (2014) Micropolar continuum modeling of bi-dimensional tetrachiral lattices. Proc R Soc A Math Phys Eng Sci 470(2165):20130734

Chu DN, Xie YM, Hira A, Steven GP (1996) Evolutionary structural optimization for problems with stiffness constraints. Finite Elem Anal Des 21(4):239–251

Courtois L, Maire E, Perez M, Rodney D, Bouaziz O, Bréchet Y (2012) Mechanical properties of monofilament entangled materials. Adv Eng Mater 14(12):1128–1133

Cui C, Ohmori H, Sasaki M (2003) Computational morphogenesis of 3D structures by extended ESO method. J Int Assoc Shell Spatial Struct 44(1):51–61

Dalaq AS, Abueidda DW, Al-Rub RKA, Jasiuk IM (2016) Finite element prediction of effective elastic properties of interpenetrating phase composites with architectured 3D sheet reinforcements. Int J Solids Struct 83:169–182

Decker L, Jeulin D, Tovena I (1998) 3D morphological analysis of the connectivity of a porous medium. Acta Stereologica 17(1):107–112

dell'Isola F, Della Corte A, Giorgio I (2017) Higher-gradient continua: the legacy of Piola, Mindlin, Sedov and Toupin and some future research perspectives. Math Mech Solids 22(4):852–872

Deshpande VS, Fleck NA (2000) Isotropic constitutive models for metallic foams. J Mech Phys Solids 48:1253–1283

Deshpande VS, Fleck NA, Ashby MF (2001) Effective properties of the octet-truss lattice material. J Mech Phys Solids 49(8):1747–1769

Dirrenberger J (2012) Effective properties of architectured materials. Ph.D. Thesis, MINESParis-Tech, Paris, Dec 2012

Dirrenberger, J.: From architectured materials to the development of large-scale additive manu-facturing. SPOOL 4(1):13–16 (2017). https://journals.library.tudelft.nl/index.php/spool/article/view/1910

Dirrenberger J, Forest S, Jeulin D (2013) Effective elastic properties of auxetic microstructures: anisotropy and structural applications. Int J Mech Mater Des 9(1):21–33. https://doi.org/10.1007/s10999-012-9192-8

Dirrenberger J, Forest S, Jeulin D (2012) Elastoplasticity of auxetic materials. Comput Mater Sci 64:57–61. https://doi.org/10.1016/j.commatsci.2012.03.036

Dirrenberger J, Forest S, Jeulin D (2014) Towards gigantic RVE sizes for stochastic fibrous networks. Int J Solids Struct 51(2):359–376. https://doi.org/10.1016/j.ijsolstr.2013.10.011

Dirrenberger J, Forest S, Jeulin D, Colin C (2011) Homogenization of periodic auxetic materi-als. Procedia Engineering 10. In: 11th international conference on the mechanical behavior of materials (ICM11), 1847–1852. https://doi.org/10.1016/j.proeng.2011.04.307

Djumas L, Molotnikov A, Simon GP, Estrin Y (2016) Enhanced mechanical performance of bio-inspired hybrid structures utilising topological interlocking geometry. Sci Rep 6:26706

Duballet R, Baverel O, Dirrenberger J (2017) Classification of building systems for concrete 3D printing. Autom Constr 83:247–258

Duballet R, Baverel O, Dirrenberger J (2018) Design of space truss based insulating walls for robotic fabrication in concrete. In: Rycke KD, Gengnagel C, Baverel O, Burry J, Mueller C, Nguyen MM, Rahm P, Thomsen MR (eds) Humanizing digital reality, Chap. 39, pp. 453-461. Springer. https://doi.org/10.1007/978-981-10-6611-5_39

Duballet R, Gosselin C, Roux P (2016) Additive manufacturing and multi-objective optimization of graded polystyrene aggregate concrete structures. In: Thomsen M, Tamke M, Gengnagel C, Faircloth B, Scheurer F (eds) Modelling behaviour- design modelling symposium 2015, Chap. Additive manufacturing and multi-objective optimization of graded polystyrene aggregate con-crete structures

Duysinx P, Bendsøe MP (1998) Topology optimization of continuum structures with local stress constraints. Int J Numer Meth Eng 43(8):1453–1478

Embury D, Bouaziz O (2010) Steel-based composites: driving forces and classifications. Annu Rev Mater Res 40:213–241

Erk KA, Dunand DC, Shull KR (2008) Titanium with controllable pore fractions by thermoreversible gelcasting of TiH2. Acta Mater 56(18):5147–5157

Eschenauer HA, Olhoff N (2001) Topology optimization of continuum structures: a review. Appl Mech Rev 54(4):331–390

Escoda J, Jeulin D, Willot F, Toulemonde C (2015) Three-dimensional morphological modeling of concrete using multiscale Poisson polyhedra. J Microsc 258(1):31–48

Fallet A, Lhuissier P, Salvo L, Bréchet Y (2008) Mechanical behaviour of metallic hollow spheres foam. Adv Eng Mater 10(9):858–862

Fleck NA, Deshpande VS, Ashby MF (2010) Micro-architectured materials: past, present and future. Proc R Soc A Math Phys Eng Sci 466(2121):2495–2516

Ford S, Despeisse M (2016) Additive manufacturing and sustainability: an exploratory study of the advantages and challenges. J Clean Prod 137:1573–1587

François D, Pineau A, Zaoui A (2012) Mechanical behaviour of materials, volume 1: microand macroscopic constitutive behaviour, vol. 180. Solid mechanics and its applications. Springer

Freeman AJ (2002) Materials by design and the exciting role of quantum computation/simulation. J Comput Appl Math 149(1):27–56

Fritzen F, Forest S, Kondo D, Böhlke T (2013) Computational homogenization of porous materials of Green type. Comput Mech 52(1):121–134

Geers MGD, Yvonnet J (2016) Multiscale modeling of microstructure-property relations. MRS Bull 41(8):610–616

Ghaedizadeh A, Shen J, Ren X, Xie YM (2016) Tuning the performance of metallic auxetic meta-materials by using buckling and plasticity. Materials 9(54):1–17

Gosselin C, Duballet R, Roux P, Gaudillière N, Dirrenberger J, Morel P (2016) Large-scale 3D printing of ultra-high performance concrete- a new processing route for architects and builders. Mater Des 100:102–109

Guest JK, Prévost JH (2006) Optimizing multifunctional materials: design of microstructures for maximized stiffness and fluid permeability. Int J Solids Struct

Gutowski TG, Branham MS, Dahmus JB, Jones AJ, Thiriez A, Sekulic DP (2009) Thermodynamic analysis of resources used in manufacturing processes. Environ Sci Technol 43(5):1584–1590

Hajela P, Kim B (2001) On the use of energy minimization for CA based analysis in elasticity. Struct Multidiscip Optim 23:24–33

Hopkins JB, Shaw LA, Weisgraber TH, Farquar GR, Harvey CD, Spadaccini CM (2016) Design of nonperiodic microarchitectured materials that achieve graded thermal expansions. J Mech Robot 8(5):051010

Hopman RK, Leamy MJ (2010) Triangular cellular automata for computing two-dimensional elastodynamic response on arbitrary domains. J Appl Mech 78(2):021020

Hor A, Saintier N, Robert C, Palin-Luc T, Morel F (2014) Statistical assessment of multiaxial HCF criteria at the grain scale. Int J Fatigue 67:151–158

Huang X, Xie YM (2007) Convergent and mesh-independent solutions for bi-directional evolutionary structural optimization method. Finite Elem Anal Des 43(14):1039–1049

Hummel RE (2004) Understanding materials science, 2 edn. Springer-Verlag, New York

Jean A, Jeulin D, Forest S, Cantournet S, N'Guyen F (2011) A multiscale microstructure model of carbon black distribution in rubber. J Microsc 241(3):243–260

Jeulin D (2000) Random texture models for material structures. Stat Comput 10(2):121–132

Jeulin D, Ostoja-Starzewski M (2001) Mechanics of random and multiscale microstructures. CISM courses. Springer, Heidelberg

Kanit T, Forest S, Galliet I, Mounoury V, Jeulin D (2003) Determination of the size of the representative volume element for random composites: statistical and numerical approach. Int J Solids Struct 40:3647–3679

Kanit T, N'Guyen F, Forest S, Jeulin D, Reed M, Singleton S (2006) Apparent and effective physical properties of heterogeneous materials: representativity of samples of two materials from food industry. Comput Methods Appl Mech Eng 195:3960–3982

Khakalo S, Niiranen J (2017) Isogeometric analysis of higher-order gradient elasticity by user elements of a commercial finite element software. Comput Aided Des 82:154–169

Khoshnevis B (2004) Automated construction by contour crafting- related robotics and information technologies. Autom Construct 13:5–19

Kim S, Abdalla MM, Gürdal Z, Jones M (2004) Multigrid accelerated cellular automata for structural design optimization: A 1-D implementation. In: 45th AIAA/ASME/ASCE/AHS/ASC structures, structural dynamics and materials conference, Palm Springs, California

Kolopp A, Rivallant S, Bouvet C (2013) Experimental study of sandwich structures as armour against medium-velocity impacts. Int J Impact Eng 61:24–35

Körner C, Liebold-Ribeiro Y (2015) A systematic approach to identify cellular auxetic materials. Smart Mater Struct 24(2):025013

Kotani M, Ikeda S (2016) Materials inspired by mathematics. Sci Technol Adv Mater 17(1):253–259

Kowalski N, Delannay L, Yan P, Remacle JF (2016) Finite element modeling of periodic polycrystalline aggregates with intergranular cracks. Int J Solids Struct 90:60–68

Labonnote N, Ronnquist A, Manum B, Rüther P (2016) Additive construction: state-of-the-art, challenges and opportunities. Autom Construct 72:347–366

Laszczyk L, Dendievel R, Bouaziz O, Bréchet Y, Parry G (2009) Design of architectured sandwich core materials using topological optimization methods. In: symposium LL-architectured multifunctional materials, vol. 1188, MRS Proceedings

Le Riche R, Haftka RT (2012) On global optimization articles in SMO. Struct Multidiscip Optim 46:627–629

Lebée A, Sab K (2012) Homogenization of thick periodic plates: application of the bending-gradient plate theory to a folded core sandwich panel. Int J Solids Struct 49(19–20):2778–2792

Lecampion B, Vanzo J, Ulm F-J, Huet B, Germay C, Khalfallah I, Dirrenberger J (2011) Evolution of portland cement mechanical properties exposed to CO2-rich fluids: investigation at different scales. In: MPPS 2011, symposium on mechanics and physics of porous solids : a tribute to Pr. Olivier Coussy

Lévy M (1874) La statique graphique et ses applications aux constructions. Gauthier-Villars, Paris

Lewandowski M, Amiot M, Perwuelz A (2012) Development and characterization of 3D nonwoven composites. In: Boudenne A (ed) Materials science forum. Polymer composite materials: From Macro, Micro to Nanoscale, vol 714, pp 131–137

Liu J, Gu T, Shan S, Kang SH, Weaver JC, Bertoldi K (2016) Harnessing buckling to design architected materials that exhibit effective negative swelling. Adv Mater 28(31):6619–6624

Madi K, Forest S, Boussuge M, Gailliègue S, Lataste E, Buffière J-Y, Bernard D, Jeulin D (2007) Finite element simulations of the deformation of fused-cast refractories based on X-ray computed tomography. Comput Mater Sci 39:224–229

Matouš K, Geers MGD, Kouznetsova VG, Gillman A (2017) A review of predictive nonlinear theories for multiscale modeling of heterogeneous materials. J Comput Phys 330:192–220

Mezeix L, Bouvet C, Huez J, Poquillon D (2009) Mechanical behavior of entangled fibers and entangled cross-linked fibers during compression. J Mater Sci 44(14):3652–3661

Michell AGM (1904) The limit of economy of material in frame structures. Phil Mag 8(6):589–597

Mirkhalaf M, Khayer Dastjerdi A, Barthelat F (2014) Overcoming the brittleness of glass through bio-inspiration and micro-architecture. Nature Commun

Missoum S, Gürdal Z, Setoodeh S (2005) Study of a new local update scheme for cellular automata in structural design. Struct Multidiscip Optim 29:103–112

Olson GB (2001) Beyond discovery: design for a new material world. Calphad 25(2):175–190

Ostoja-Starzewski M (2008) Microstructural randomness and scaling in mechanics of materials. Mordern mechanics and mathematics. Chapman & Hall/CRC

Peyrega C, Jeulin D, Delisée C, Malvestio J (2011) 3D morphological characterization of phonic insulation fibrous media. Adv Eng Mater 13(3):156–164

Placidi L, Barchiesi E, Della Corte A (2017) Identification of two-dimensional pantographic structures with a linear d4 orthotropic second gradient elastic model accounting for external bulk double forces. In: dell'Isola F, Sofonea M, Steigmann D (eds) Mathematical modelling in solid mechanics, Advanced structured materials, Chap 14, vol 69, pp 211–232. Springer, Singapore. https://doi.org/10.1007/978-981-10-3764-1_14

Placidi L, El Dhaba AR (2015) Semi-inverse method à la Saint-Venant for two-dimensional linear isotropic homogeneous second-gradient elasticity. Math Mech Solids 22(5):919–937

Rayneau-Kirkhope D, Mao Y, Farr R (2012) Ultralight fractal structures from hollow tubes. Phys Rev Lett 109(204301)

Ren X, Shen J, Ghaedizadeh A, Tian H, Xie YM (2016) A simple auxetic tubular structure with tuneable mechanical properties. Smart Mater Struct

Rosi G, Auffray N (2016) Anisotropic and dispersive wave propagation within strain-gradient framework. Wave Motion 63:120–134

Rozvany GIN (2009) A critical review of established methods in structural topology optimization. Struct Multidiscip Optim 37:217–237

Rozvany GIN, Bendsøe MP, Kirsch U (1995) Layout optimization of structures. Appl Mech Rev 48(2):41–119

Salonitis K, Chantzis D, Kappatos V (2017) A hybrid finite element analysis and evolutionary computation method for the design of lightweight lattice components with optimized strutdiameter. Int J Adv Manufact Technol 90(9–12):2689–2701

Schaedler TA, Carter WB (2016) Architected cellular materials. Ann Rev Mater Res 46:187–210

Schaedler TA, Jacobsen AJ, Torrents A, Sorensen AE, Lian J, Greer JR, Valdevit L, Carter WB (2011) Ultralight metallic microlattices. Science 334(6058):962–965

Schindler S, Mergheim J, Zimmermann M, Aurich JC, Steinmann P (2017) Numerical homogenization of elastic and thermal material properties for metal matrix composites (MMC). Continuum Mech Thermodyn 29(1):51–75

Shim J, Shan S, Košmrlj A, Kang SH, Chen ER, Weaver JC, Bertoldi K (2013) Harnessing instabilities for design of soft reconfigurable auxetic/chiral materials. Soft Matter 9(34):8198–8202

Sigmund O (2011) On the usefulness of non-gradient approaches in topology optimization. Struct Multidiscip Optim 43:589–596

Smith P, Torquato S (1988) Computer simulation results for the two-point probability function of composite media. J Comput Phys 76(1):176–191

Steeves CA, Santos e Lucato SL, dos He M, Antinucci E, Hutchinson JW, Evans AG (2007) Concepts for structurally robust materials that combine low thermal expansion with high stiffness. J Mech Phys Solids 55:1803–1822

Stoychev G, Zakharchenko S, Turcaud S, Dunlop JWC, Ionov L (2012) Shape-programmed folding of stimuli-responsive polymer bilayers. ACS Nano 6(5):3925–3934

Torabian N, Favier V, Dirrenberger J, Adamski F, Ziaei-Rad S, Ranc N (2017) Correlation of the high and very high cycle fatigue response of ferrite based steels with strain ratetemperature conditions. Acta Mater 134:40–52

Torabian N, Favier V, Ziaei-Rad S, Adamski F, Dirrenberger J, Ranc N (2016) Self-heating measurements for a dual-phase steel under ultrasonic fatigue loading for stress amplitudes below the conventional fatigue limit. Proc Struct Integr 2:1191–1198

Torabian N, Favier V, Ziaei-Rad S, Dirrenberger J, Adamski F, Ranc N (2017) Calorimetric studies and self-heating measurements for a dual-phase steel under ultrasonic fatigue loading. In: Wei Z, Nikbin K, McKeighan P, Harlow D (eds) Fatigue and fracture test planning, test data acquisitions and analysis, vol STP1598, pp 81–93, ASTM (2017). https://doi.org/10.1520/STP159820160053

Torabian N, Favier V, Ziaei-Rad S, Dirrenberger J, Adamski F, Ranc N (2016) Thermal response of DP600 dual-phase steel under ultrasonic fatigue loading. Mater Sci Eng A Struct Mater Prop Microstruct Process 677:97–105

Torquato S (1998) Morphology and effective properties of disordered heterogeneous media. Int J Solids Struct 35(19):2385–2406

Torquato S (2001) Random heterogeneous materials. Springer

Tovar A, Niebur GL, Sen M, Renaud JE, Sanders B (2004) Bone structure adaptation as a cellular automaton optimization process. In: 45th AIAA/ASME/ASCE/AHS/ASC Structures, structural dynamics & materials conference, Palm Springs, California

Tovar A, Patel NM, Kaushik AK, Renaud JE (2007) Optimality conditions of the hybrid cellular automata for structural optimization. AIAA J 45(3):673–683

Trinh DK, Jänicke R, Auffray N, Diebels S, Forest S (2012) Evaluation of generalized continuum substitution models for heterogeneous materials. Int J Multiscale Comput Eng 10(6):527–549

Turcaud S, Guiducci L, Fratzl P, Dunlop JWC, Bréchet Y (2011) An excursion into the design space of biomimetic architectured biphasic actuators. Int J Mater Res 102(6):607–612

Vicente WM, Zuo ZH, Pavanello R, Calixto TKL, Picelli R, Xie YM (2016) Concurrent topology optimization for minimizing frequency responses of two-level hierarchical structures. Comput Methods Appl Mech Eng 301:116–136

Wang ZP, Poh LH, Dirrenberger J, Zhu Y, Forest S (2017) Isogeometric shape optimization of smoothed petal auxetic structures via computational periodic homogenization. Comput Methods Appl Mech Eng 323:250–271

Weaver PM, Ashby MF (1996) The optimal selection of material and section-shape. J Eng Des 7(2):129–150

Xie YM, Steven GP (1993) A simple evolutionary procedure for structural optimization. Comput Struct 49:885–896

Xu B, Huang X, Zhou SW, Xie YM (2016) Concurrent topological design of composite thermoelastic macrostructure and microstructure with multi-phase material for maximum stiffness. Compos Struct 150:84–102

Xu S, Shen J, Zhou S, Huang X, Xie YM (2016) Design of lattice structures with controlled anisotropy. Materials and Design

Yeong CLY, Torquato S (1998) Reconstructing random media. Phys Rev E

Yoon HS, Lee JY, Kim HS, Kim MS, Kim ES, Shin YJ, Chu WS, Ahn SH (2014) A comparison of energy consumption in bulk forming, subtractive, and additive processes: review and case study. Int J Precis Eng Manufact Green Technol 1(3):261–279

Zakhama R, Abdalla MM, Smaoui H, Gürdal Z (2009) Multigrid implementation of cellular automata for topology optimization of continuum structures. Comput Modeling Eng Sci 51(1):1–24

Zhou S, Li Q (2008) Design of graded two-phase microstructures for tailored elasticity gradients. J Mater Sci 43:5157–5167

Chapter 5
Robotic Building as Integration of Design-to-Robotic-Production and -Operation

Henriette Bier⊕, Alexander Liu Cheng, Sina Mostafavi, Ana Anton and Serban Bodea

Abstract *Robotic Building* implies both physically built robotic environments and robotically supported building processes. Physically built robotic environments consist of reconfigurable, adaptive systems incorporating sensor-actuator mechanisms that enable buildings to interact with their users and surroundings in real-time. These robotic environments require *Design-to-Production and -Operation* (D2P&O) chains that may be (partially or completely) robotically driven. This chapter describes previous work aiming to integrate D2RP&O processes by linking performance-driven design with robotic production and user-driven building operation.

5.1 Introduction

While architecture and architectural production are increasingly incorporating aspects of non-human agency employing data, information, and knowledge contained within the (worldwide) network connecting electronic devices, the question is not *whether* but *how* robotic systems can be incorporated into building processes and buildings (Oosterhuis and Bier 2013). This chapter aims to answer this question by reflecting on the achievements of the *Robotic Building* (RB) team at *Technical University Delft* (TU Delft) and by identifying future steps. The focus is on an architecture that is robotically enabled to interact with its users and surroundings in real-time and the corresponding *Design-to-Production and -Operation* (D2P&O) processes that are (in part or as whole) robotically driven. Such modes of production and operation involve agency of both humans and non-humans. Thus agency is not located in one or another but in the heterogeneous associations between them (Latour 2009).

H. Bier (✉) · A. Liu Cheng · S. Mostafavi · A. Anton · S. Bodea
Faculty of Architecture and the Built Environment, TU Delft, Delft, The Netherlands
e-mail: h.h.bier@tudelft.nl

H. Bier · S. Mostafavi
Dessau Institute of Architecture, HS Anhalt, Dessau, Germany

© Springer International Publishing AG, part of Springer Nature 2018
H. Bier (ed.), *Robotic Building*, Springer Series in Adaptive Environments,
https://doi.org/10.1007/978-3-319-70866-9_5

This chapter describes attempts to integrate *Design-to-Robotic-Production* (D2RP) with *Design-to-Robotic-Operation* (D2RO) processes by linking design and production with smart operation of the built environment and by advancing applications in performance optimization, robotic manufacturing, and user-driven building operation.

5.2 Robotic Building

RB relies on interactions between human and non-human or cyber-physical agents not only at design and production level but also at building operation level, wherein users and environmental conditions contribute to the emergence of various architectural configurations. Such physically built robotic environments incorporate sensor-actuator mechanisms that enable buildings to interact with their users and surroundings in real-time (see Fig. 5.1). Their conceptualization and materialization require D2RP&O processes that link design to production and building operation (Fig. 5.2). In this context, design becomes process-instead of object-oriented and use of space becomes time-instead of program-or function-based. This implies that architects increasingly design processes, while users operate multiple time-based architectural configurations (Bier and Knight 2014) emerging from the same physical space that may physically or sensorially reconfigure in accordance to environmental and user-specific needs.

In this context, spatial and ambiental reconfiguration optimises use of space by facilitating changing uses of physically built space within reduced timeframes (Liu Cheng and Bier 2016a, b). Furthermore, it reduces energy consumption by employing passive and active climate control and ensures local ambient customisation. Such spatial and ambiental reconfiguration requires virtual modelling and simulation that interface the production and real-time operation of physically built space (Bier and Knight 2014), thus establishing an unprecedented D2RP&O feedback loop, which is the focus of this chapter.

5.2.1 Design-to-Robotic-Production

Industrial robots have been used in a wide range of production processes since the 70s but only more recently academia and creative industry started to explore their potential in architecture. More than 90 institutions and start-ups employ today industrial robots for either developing 1:1 prototypes of bare structures or building components[1] that are integrated in buildings designed and constructed conventionally. In contrast, D2RP aims to introduce strategies for the integral production of buildings

---

[1] The Robotics in Architecture map (accessed from http://www.robotsinarchitecture.org/map-of-cr eative-robots) shows that more than 90 institutions and start-ups are using robots worldwide.

addressing all structural, environmental, climatic, programmatic, and user-specific, etc. needs. This implies that the complete building process is taken in consideration in order to identify requirements for the robotic production. The goal is to integrate production aspects from the early stages of design.

Several experiments with optimized additive and subtractive production of computationally derived architectural and structural topologies have been implemented at scales ranging from architectural (*macro*) to componential (*meso*) and material (*micro*) scale. By linking performance-based and generative design methods to robotic manufacturing, D2RP processes establish a feedback-loop between design and production of buildings components at full-scale.

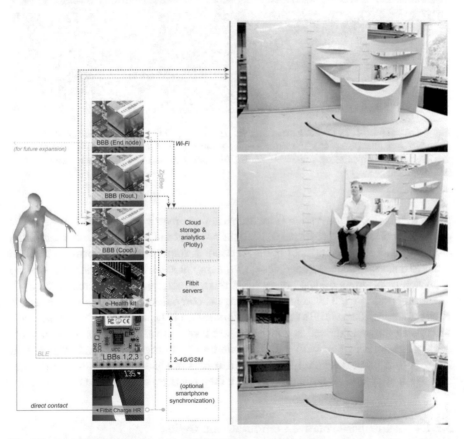

Fig. 5.1 *Design-to-Robotic-Operation* (D2RO) links computational mechanisms and services to spatial reconfiguration for the promotion of occupant well-being (Liu Cheng and Beir 2016a, b). Left: Basic form of the System Architecture, illustrating the relationship between (1) the Local System, (2) the Wearables Subsystem, and (3) the Cloud/Remote Services Subsystem, which are conceived as the essential features of D2RO. Right: *Proof-of-concept* prototype whose physical transformations actuate in response to sensed and processed data

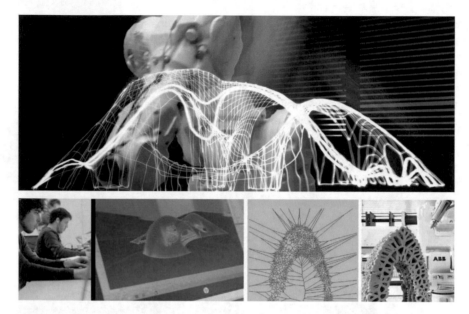

Fig. 5.2 Design-to-Robotic-Production establishing a direct link between virtual modelling and physical fabrication (2014–16). The virtual model (bottom-left) is translated into robotic paths (top) that are further refined using structural analysis in order to robotically produce a clay prototype (bottom-right)

D2RP involves a conversion from the virtual geometric model, which is often the result of optimization processes (e.g. functional, formal, structural, environmental, etc.), into suitable robotic tool paths to deposit, remove, or transform material in order to materialize the intended design. At a digital level, a parametric form-finding approach involving amongst others functional, structural, and environmental optimization is adopted. This approach relies on computational methods such as the *Finite Element Method* (FEM), *Computational Fluid Dynamics* (CFD), etc. Furthermore, material and fabrication constraints are taken into account in order to connect physical materialization with virtual modelling and simulation. This implies that multi-performative design relying on multi-robots production and multi-scale materialisation integrates all requirements from the very beginning of the D2RP process.

5.2.1.1 Multi-performative Computational Design

Architecture is typically developed and built at several discrete scales. While the multi-scalar approach has been the subject of research and debate across architectural history, only more recently—and due to advances in modelling, simulation, and robotic technology—architecture adopted a real-scale design paradigm. In this context, design-to-production focused computation embeds the versatility of computational design into fabrication processes that are accessible to the industry and designer community. Such integration of computational design into production processes optimizes fabrication resolution, enables novel designs, and promotes a holistic approach in architecture.

Computational design methods developed for D2RP largely rely on recursive computation where once produced geometric results are propagated across design and fabrication iterations resulting in the development of an unified multi-scalar production approach. Consequently, traditional indications of scale—from detail to assembly—and architectural space are translated into wider ranges (*micro*, *meso* and *macro*) that operate as bounds indicative of suitability of particular fabrication techniques and respective recursive depths.

D2RP establishes a feedback-loop between design and fabrication by linking design and simulation environments—e.g., Rhinoceros and Grasshopper—to robotic manufacturing. The role of computation in such robotic production systems is extended, firstly, by the way machines are programmed and, secondly, by the way materials are processed. In the *recursive milling* case study (see Fig. 5.3), continuous robotic paths with embedded information pertaining to material and fabrication constraints generate overall form and surface-texture. The optimised path is a self-avoiding curve[2] that translates into a minimum-length tool-path, featuring low- and high-resolution, for fast and slow material removal.

Since it is particularly suited for delivering designs across multiple scales, recursive milling informs not only subtractive but also additive D2RP. The technology allows access to and control over the internal structure of an object, making the interior of the geometry to design an important subject of research. Variable porosity embodying quantifiable relations between matter and void are employed within D2RP in order to improve environmental performance of building components and reduce material usage. In this context, robotic path-constraints are employed as design drivers to create informed tectonics at volumetric and surface texture levels (see Figs. 5.2 and 5.4). The robotic motion defines the boundaries of the digital design-space in relation to the physical solution-space informing the parametric setup with

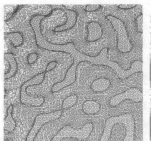

Fig. 5.3 Recursive milling method with homogenous resolution (left), tool path with informed resolution based on material removal (middle), and prototyping (right)

---

[2]For example, a Hilbert space-filling curve, which was first described by mathematician David Hilbert in 1891.

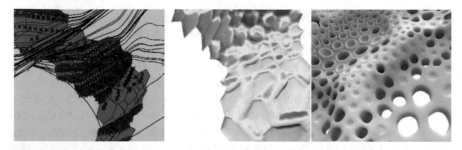

Fig. 5.4 Materialization of informed porosity using structural (left) and environmental analysis (middle) is computed for volumetric tectonics and surface textures

respect to ranges of reachability and optimum tool orientations, thus contributing to enlarging the solution-space.

Structural optimization for additive D2RP, involves methods for form finding of compression-only structures, derived from the innate characteristics of high viscosity ceramic clays. While the optimization takes local and global load and support conditions into consideration, at the *macro* level, a compression-only structure is developed, whose porosity at this scale fulfils functional and aesthetic requirements (Fig. 5.2). At the *micro* level, in order to achieve material porosity, a finite element method for optimizing material distribution is used on selected fragments of the structure (Fig. 5.4). Various algorithmic form finding and optimization techniques, mostly in Rhinoceros–Grasshopper and Python are applied in order to enable the systematic exploration and evaluation of design alternatives within the design-solution space, eventually providing the required information for production.

5.2.1.2   Multi-mode and -Robot Production

As part of a larger D2RP&O framework, D2RP is aiming at integrating the design, fabrication, and operation of buildings in order to address the increasing interest of the construction industry in automation at both production and building operation level. At its core the D2RP system has a cyber-physical setup wherein fabrication sequences are informed by design iterations and simulated kinematic processes. This integrated D2RP approach meeting various demands of the built environment is oriented towards informing building construction processes.

In the workshop at *InDeSem 2015* organised at TU Delft, a compact multi-mode and -robot production setup—consisting of three industrial robots equipped with various tools—was installed in one day to address a large array of manufacturing tasks. Most importantly, these industrial robots were linked directly to computational design environments (Mostafavi and Bier 2016). Once this connection was established, even users at beginner level—e.g., students that have never programmed an industrial robot before—were able to effectively asses fabricability of their designs and optimize the iterations for milling and hot-wire cutting of *Expanded Polystyrene*

(EPS) foam (Mostafavi et al. 2015). Volumetric cutting was used for material removal and general shaping of components while milling was used for adding surface texture and controlling porosity. Such a multi-robot production setup—a de facto small production line—ensures increased efficiency of production while relying on interaction between human and robot agents.

In addition to subtractive D2RP, additive methods were explored where the reach and reduced weight of industrial robots in the small-medium range makes the easily adjustable production unit perfect for the production of small-medium building components. Furthermore, self-developed end-effectors were used for best results in the controlled deposition of customized materials according to patterns that resulted from the structural and robotic path optimization routines (see Fig. 5.5). In this context, the innovation lies in printing with customized materials and end effectors on customized substrates. The robotic setup is flexible enough to allow for the programming of custom paths so that previously fabricated EPS substrates can be used to produce flat or curved ceramic clay pieces (Pottmann et al. 2012). This 3D printing technique is reaching *Technology Readiness Level* (TRL) 6 and could be tested now in an operational environment.

5.2.1.3 Multi-scale Materialization

D2RP employs various materials and relies on multiple robotic production methods in order to achieve quantifiable design performances. Until now materiality, as interface between digital design space and physical fabrication, has been mainly defined along three performance criteria: spatial functionality, structural strength, and environmental efficiency. Furthermore, by integrating computation and robotic materialization, D2RP introduces strategies for extending the design space. With computation implemented at multiple scales and with multi-robot setups enhanced by multi-mode techniques the design space is enlarged. Such multi-mode techniques

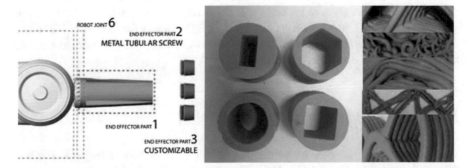

Fig. 5.5 Customized 3D printed end-effector and materials for robotic additive manufacturing using ceramic clay

Fig. 5.6 Customized robotic setup for 3D printing of optimized ceramic clay patterns on ruled (left) and flat (right) surfaces

may involve hybrid production approaches that integrate multiple methods of processing materials as for instance, subtractive, additive, and formative.

Considering the building scale, assembly methods allowing expansion beyond the size of building components, which are limited by the actual size of the production space, enlarge the design space as well. These may involve material handling i.e. feeding the components to the robot, picking/gripping and assembling/joining components, while using force control and control of chained tolerances, etc. If multi-robots and -modes processes have been explored since 2014, assembly methods still need to be developed. Thus, multi-scale materialization scenarios, wherein different manufacturing and assembly operations are combined need to be now explored and advanced in order to push the bare structural prototype towards becoming a building.

If the until now developed multi-performative, -mode, and -robot D2RP reaches TRL levels ranging between 4 and 6, multi-scale D2RP is still at its very beginning (Fig. 5.6). Robotisation in building construction by translating building and material performances from discretized geometry into continuous optimized robotic paths (for material deposition, subtraction, or transformation) and by developing coordination scenarios for multi-robot operations in order to involve several robots in the process of production either simultaneously or in short sequence still requires research. In particular the integration between D2RP and D2RO requires further definition, since D2RP pursues robotisation in building construction, while D2RO aims to achieve robotisation in the operation of buildings.

5.2.2 Design-to-Robotic-Operation

Discussions on *intelligence integrated into the built-environment* began in the late 60 s and early 70s (Cook 1970, 1972; Eastman 1972; Pask 1975a, b; Negroponte 1947, 1975). They belonged to a broader discourse that engaged various domains and disciplines in the exploration of opportunities entailed by the *Information Age*. During this period, and partly due to the novelty of the exploration as well as to the rudimentary state and forbidding costs of *Information and Communication Technologies* (ICTs), these discussions were principally theoretical and/or hypothetical in nature. Over the next two decades, the discourse specialized into subset fields broadly coalescing into the technical on the one hand and the architectural on the other.

With respect to the technical, *Ambient Intelligence* (AmI) was coined in the late 90 s to describe a cohesive vision of a future *digital living room*, a built-environment whose computing hardware and software technology imbued its dwelling space with serviceable intelligence to the benefit of its occupant(s) (Zelkha et al. 1998). Within AmI a further specialized domain developed, i.e., that of *Ambient Assisted Living*—or *Active and Assisted Living*, as preferred by the European Union—(AAL), which framed its inquiry around the promotion of quality of life as well as the prolongation of independence with respect to *Activities of Daily Living* (ADLs) among the elderly via technical assistance. By the first decade of the 21st century, AmI and AAL were established and proliferating topics within the fields of Computer Science and related Engineerings (Augusto et al. 2010; Esch 2013; Cook et al. 2009; Nakashima and Aghajan 2010), Architectural Engineering (Bock et al. 2015; Georgoulas et al. 2014; Linner et al. 2012), and—indirectly—in the Medical Sciences (Acampora et al. 2013).

With respect to the architectural, and beginning with Price's pioneering *Generator Project* and corresponding programs by Frazer and Frazer (1979) in the late 70s, notions of interaction between human and non-human agents in the built-environment began to be explored. For example, in Price's project, architecture was conceived as a set of interchangeable subsystems integrated into a unifying computer system, which enabled a reconfigurability sensitive to function. More importantly, both Price and Frazer intended for the system itself to suggest its own reconfigurations,[3] denoting non-human agency in the built-environment. Although the *Generator Project* was never realized, it became the de facto first instance of a subset field in Architecture concerned with bi-directional communication and interaction between human and non-human agents in the built-environment, viz., *Interactive Architecture* (IA) (Fox and Kemp 2009; Fox 2010; Oosterhuis 2012) first and *Adaptive Architecture* (AA) (Jaskiewicz 2013; Kolarevic 2014; Schnädelbach 2010) later, which—like AmI—have also proliferated in the 21st century.

The embedding of intelligence into the built-environment with respect to AmI/AAL and to IA/AA has differed in sophistication, with the former far surpassing the latter in terms of technical complexity, reliance, and performance. This has been largely due to their differing emphases, with the technical focusing on computing hardware and software technology and the architectural on spatial experience, materiality, and form. That is, the technical proliferated with resources resulting

---

[3] Steenson quotes (2014) two interesting excerpts from letters exchanged by Price and the Frazers. First, from Price to the Frazers, stating his objective: "*The whole intention of the project is to create an architecture sufficiently responsive to the making of a change of mind constructively pleasurable*" (Price et al. 1978). Second, from the Frazers to Price, expressing a desired characteristic: "*If you kick a system, the very least that you would expect it to do is kick you back*" (Frazer and Frazer 1979).

from robust and sustained computational development over decades in ways that the architectural could not, at least not with the same affinity and immediacy. Nevertheless, technical sophistication or lack thereof alone has not necessarily guaranteed or disqualified contributions in the discourse. Indeed, principally technical as well as principally architectural explorations have both independently identified key *effective* as well as *affective* desiderata common to built-environments—intelligent or otherwise—construed as successful with respect to function as well as to spatial experience. This consideration, however, includes a caveat: while both the technical as well as the architectural have yielded independent contributions, these have been otherwise limited by the lack of mutually provided input and/or feedback. For example, AmI/AAL may continue to proliferate as a technical subject even if the physical aspect of its built context remains presupposed and/or static to conventional design and construction frameworks. Similarly, IA/AA may also continue to proliferate in its affective and/or qualitative explorations even if the technical aspects of its implementations express modest computational sophistication. However, the promise of solutions yielded by both principally technical AmI/AAL and principally architectural IA/AA explorations will be unwittingly and invariably limited by the rigid and increasingly outdated character of their complementing frameworks. This is because the sophistication of a system will depend on that of its mutually complementing subsystems; and two or more subsystems may not mutually complement, sustain, and/or support one another properly if their levels of development and sophistication do not correspond (Milgrom 1990). More succinctly expressed: at present, the architectural does not correspond to the technically superior AmI/AAL, while the technical does not correspond to the architecturally superior IA/AA. Consequently, a different design paradigm/framework is required in order to enable comprehensively and cohesively intelligent built-environments with corresponding levels of technical and architectural sophistication.

In this section, principles and strategies developed at TU Delft are introduced as *Design-to-Robotic-Operation* (D2RO), which is presented and promoted as part of an alternative design and development paradigm (i.e., D2RP&O) of intelligent built-environments that considers the technical as well as the architectural in conjunction from the early stages of the design and development processes. In this manner, the built-environment is construed as a highly sophisticated and integrated *Cyber-Physical System* (CPS) (Rajkumar et al. 2010) consisting of mutually informing computational and physical mechanisms that operate cooperatively and continuously via a highly heterogeneous, partially meshed, and self-healing *Wireless Sensor and Actuator Network* (WSAN) (Yang 2014). Via a series of limited and progressively complex proof-of-concept implementations (Liu Cheng 2016, Liu Cheng and Bier 2016a, b; Liu Cheng et al. 2017; Liu Cheng et al. 2017), the feasibility and promise of D2RO are demonstrated and validated.

The current state and development of D2RO is described in the following seven subsections, the first corresponding to the underlying and enabling ICT framework, and the remaining six to mechanisms and/or features that—in conjunction—service an intelligent built-environment capable of intuitive action, reaction, and interaction as well as proactive intervention. The subsystems detailed in these sections have been implemented from medium to high TRLs (i.e., 5–9) (European Association of Research and Technology Organisations (EARTO) 2015), and constitute an architecturally limited yet technically integrated whole with an overall system TRL of 5.[4] Accordingly, and while those subsystems with TRL 9 are ready to be deployed within commercial solutions, the overall system continues to be developed further both to higher degrees of TRL as well as to include additional subsystems to expand its capabilities.

5.2.2.1   System Architecture

This system at TRL 9 level consists of the following four subsystems (see Fig. 5.7): (1) a *Local system*, which establishes the WSAN; (2) a set of *Wearables*, which extend network's sensing capabilities to include more personal ranges; (3) *Remote/Cloud Services*, which connect the network with Internet-based services and unctions; and (4) Ad Hoc *Support* interfaces, which enable direct user-interventions within the network.

The main difference of the present architecture from that of existing AmI frameworks/solutions is that its functions are not centered on a locally *structured environment*. Instead, the present system is a subsystem within a larger whole. It is *extended* in terms of both its sensing as well as its actuation capabilities, both of which may perform beyond the local *structured environment*. For example—and with respect to *sensing*—in the present architecture, the *local system* continues to monitor the user's activity levels even when he/she is outside of the local *structured environment*. That is, the user-activity recorded by an activity tracker (see item 9, Fig. 5.7) is downloaded by the *local system* from the tracker's manufacturer's servers via an official *Application Program Interface* (API). This enables the local WSAN to process user-activity data continuously, which is necessary in order to develop high-fidelity personalization (Liu Cheng and Bier 2016a, b). With respect to *actuating*, in a situation where the user has collapsed and is unresponsive, the system is capable of acting beyond its local *structured environment* by sending free as well as fee-based SMS/email notifications to care-takers and/or family-members for intervention purposes (Liu Cheng et al. 2016).

Another difference is that the underlying and enabling WSAN is designed as highly heterogeneous—in terms of hardware, software, and communication protocols—in order to subsume functional, operational, and economic advantages across technologies (see Fig. 5.7). Admittedly, researchers have noted that commercial and/or proprietary solutions are often closed, rendering seamless integration with non-commercial and/or non-proprietary solutions highly cumbersome (at best) or unfeasible (at worst) (Harrison et al. 2010). This has raised challenges related to interoperability within heterogeneous systems (see Jiménez-Fernández et al. (2013),

---

[4]For reasons pertaining to system reliability and robustness, the overall TRL is determined by the least developed subsystem, as the failure of subsystem may compromise the serviceability and performance of the whole.

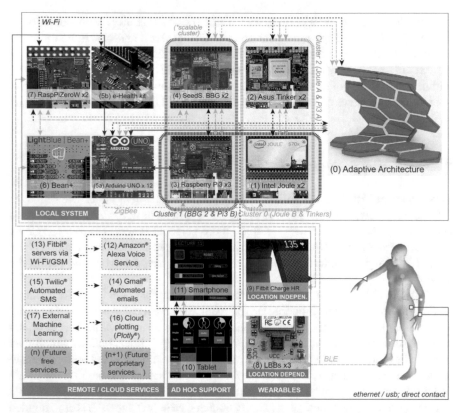

Fig. 5.7 Present state of the *Design-to-Robotic-Operation* (D2RO) System Architecture shown in its basic form, in Fig. 5.1, *Left*. This System Architecture adds a fourth subsystem to the previously identified three subsystems: (1) the Local System; (2) the Wearables Subsystem; (3) the Remote/Cloud Services Subsystem; and (4) the Ad Hoc Support Subsystem

for example), which is partly the reason why some AmI solutions have implemented homogeneous products and/or protocols. Nevertheless, in the last five years manufacturers of proprietary products and services have acted on a vested interest in making their products interoperable with a variety of systems in order to broaden their market. Consequently, an increasing number of proprietary APIs have enabled seamless integration of some proprietary products and services with non-proprietary counterparts.

By virtue of its framework of subsystems as well as of its heterogeneity, the system is highly scalable and open, capable of growing or shrinking to fit a variety of scales and scopes; and of integrating newer devices and of deprecating outdated ones in order to respond more appropriately to evolving tasks at hand.

Subsystem 1: Underlying Mechanism, Local System

This system represents the core of the WSAN. In it a variety of *Microcontroller Units* (MCUs) and development platforms serve as nodes dependent on the local structured environment. Nodes with low-storage and limited information processing capabilities serve as low-energy end devices/routers, and are principally responsible for intermittent sensor-data gathering and relaying.[5] These nodes communicate via BLE in low-range and ZigBee in high-range. Nodes with open storage-capacities, medium-performance information processing capabilities gather and store raw sensor data, parse it, and both make it available to any nodes in the network as well as stream it to Plotly® via WiFi.[6]

Nodes with high-performance information processing capabilities are principally responsible for coordination and computation.[7] These nodes may be clustered to form more powerful nodes depending on the load-requirement and exchange data with one another and with other nodes via WiFi, BLE, or ZigBee depending on the frequency as well as the latency-requirement. In one particular case, wired connections are used between nodes for data exchange (i.e., item 5 with 3, Fig. 5.7). If necessary, all Linux-running devices, regardless of individual computational power or predetermined function, may conform a cluster.

The present configuration is one of possible many. The items featured as well as the multiple instances of each serve to represent a typical highly heterogenous (both in terms of architecture as well as communication protocols and services) and cost-effective foundation capable of sustaining the growing complexity of subsequent developments and implementations.

Subsystem 2: Wearable Devices

A set of three Light Blue Bean™s (LBBs) conform the location dependent wearables while a Fitbit® Charge HR™ activity tracker (item 9, Fig. 5.7) the location independent wearable. The former detects movement in the upper-body, upper- and lower-extremities and advises the system to listen for *Open Sound Control* (OSC) packets corresponding to accelerometer data sent from a smartphone (see subsystem 3 below). Alternatively, if no smartphone is present, the LBBs broadcast accelerometer data via BLE into the system as well. This alternative is relegated to a contingency measure due to the energy-consumption of constant and sustained data streaming. Both OSC and BLE accelerometer data are used to build and update *Support Vector Machine* (SVM) and k-*Nearest Neighbor* (k-NN) classification models and to feed real-time data in the *Machine Learning* (ML) mechanism for *Human Activity Recognition* (HAR).

---

[5] Viz., PunchThrough® Bean+™ and Arduino® UNO™—items 5, 6, and 8, Fig. 5.7.

[6] Viz., Raspberry® Pi Zero W™ (RPiZW)—item 7, Fig. 5.7.

[7] Viz., Intel® Joule™, Asus® Tinkerboard™, Raspberry® Pi 3™ (RPi3) and SeedStudio® BeagleBone Green™—items 1–4, Fig. 5.7.

The principal function of the activity tracker is to gather heart-rate and physical activity (in terms of steps taken and distance covered) data continuously regardless of the location. When the user is inside the structured environment, the LBBs in conjunction with a smartphone also provide user-activity data to the system for HAR. But when the user is outside of the environment, the WSAN continues to draw limited data gathered by the activity tracker by downloading it from Fitbit®'s servers (the tracker synchronizes with the servers via mobile data when WiFi is unavailable).

Subsystem 3: Remote/ Cloud Services

Six cloud-based services conform this subsystem, three of which were first integrated in the ISARC 2016 conference article (Liu Cheng and Bier 2016a, b), and three others newly integrated into the current ecosystem. The inherited three are the following: (I) external ML mechanism via MATLAB® (item 17, Fig. 5.7); (II) data exchange with Fitbit®'s servers via its API (item 13, Fig. 5.7) and (III) cloud data-storage and -plotting via Plotly®'s API (item 16, Fig. 5.7). And the newly integrated three are the following: (IV) Amazon®'s AVS (item 12, Fig. 5.7); (V) automated SMS notifications, both via Twilio®'s API (item 15, Fig. 5.7) as well as via a T35 GSM shield as part of one of the end-device nodes of subsystem 1; and (VI) automated email notifications via Gmail©'s API (item 14, Fig. 5.7).

Subsystem 4: Ad Hoc Support Devices

In the last five years, smartphones have become convenient and ubiquitous tools for the tracking of inhabitants across a space (Andò et al. 2014), fall detection (Liu Cheng et al. 2016; Abbate et al. 2012), and HAR via ML (Anguita et al. 2013; Ortiz 2015; Micucci et al. 2017), which in conjunction with their battery life and rechargeability are the principal reasons why it they are the preferred means of accelerometer-data gathering in this development. In addition to this function, a user-interface/configuration mechanism is also enabled via a proprietary (viz., TouchOSC™ by Hexler Limited®) and a free (viz., Control by Charlie Roberts) smartphone application. This mechanism enables the user to override automation by permitting manual input/configuration.

Similarly, a tablet device has also been integrated into the ecosystem in order to provide both another user interface with a more comfortable viewing area as well as a means to modify the behavior of the LBBs and Bean+devices via BLE. Unlike the Linux-based devices of the ecosystem, the LBBs and Bean+cannot be accessed wirelessly via *Secure Shell* (SSH). Nevertheless, any necessary modifications to the devices' program or sketch may be effected wirelessly via the tablet. For example, one of the LBBs could be tasked with gathering temperature data on the user for a certain period of time and at varying intervals instead of notifying acceleration events. Both the smartphone and the tablet may access the LBBs and Bean+devices via BLE, and both are installed with the user-interface/configuration applications to enable parallel modifications should this be necessary.

5.2.2.2 Global/Local Ventilation Mechanism

This mechanism reaching TRL 5 is first implemented and tested via an abstracted surrogate model equipped with twelve DHT-22 temperature and humidity sensors, twelve air-quality sensors,[8] and twelve small DC-motor fans connected to three RPiZWs and one RPi3.

As corroborated by the *Comité Européen de Normalisation* (CEN) Standard EN15251-2007 (2007) as well as ASHRAE Standard 55-2013 and Standard 62.1-2013 (2013), the thermal Environmental Conditions for Human Occupancy with respect to comfort should be 67 to 82 °F. (~19.5–27.8 °C.) (ASHRAE® Standard 2013), while relative humidity in occupied spaces be less than 65% in order to discourage microbial growth. Furthermore, independent of human comfort considerations, frequent and consistent ventilation reduces the concentration of toxins in the air as well as the prevalence of airborne diseases (2009). In this *proof-of-concept* setup, if the collective temperature or humidity levels exceed recommended limits for comfort, all the fans activate, thereby drawing fresh air into the inhabited space (i.e., Global ventilation concept). If, however, certain areas exceed either or both limits, only those fans within and surrounding them activate (i.e., Local ventilation concept). The same concept holds for instances of air-pollution.

5.2.2.3 Voice-Control Mechanism via *Alexa Voice Service*

This mechanism reaching TRL 9 is implemented and tested via the same RPi3 mentioned in the previous section, an open-source repository using Amazon®'s API (GitHub Inc.© 2017), and a generic microphone as well as repurposed speakers. The flexibility of developing custom—and more affordable—*Alexa-enabled Devices* permits virtually any built-environment device, whether deployed in an architectural or an urban context, to capitalize from AVS.

Two main objectives inform the present integration. The first is to enable a powerful and scalable voice-control mechanism within the present development. The second is to demonstrate a cohesive technological heterogeneity between an open-source WSAN and a proprietary commercial service without additional cost (with respect to Fitbit® and Gmail©) or with minimum cost. This latter consideration connects a local intelligent-built environment with vast resources in the WWW, enabling the user to engage in a variety of activities from streaming music to purchasing groceries via devices fundamentally embedded into the built-environment.

---

[8]Viz., three of each: MQ-3 *Alcohol*, MQ-4 *Methane*, MQ-7 *Carbon Monoxide*, and MQ-8 *Hydrogen Gas*.

In the present state of D2RO, the scope of service of AVS is limited to predefined web-based skills. Work is being undertaken to expand scope to encompass services deployable within the local structured environment by either integrating a growing number of smart-home products compatible with AVS or by creating custom skills to suit specific *Internet of Things* (IoT) open-source devices via ASK.

5.2.2.4   Intervention via SMS and Email Notifications Mechanism[9]

This mechanism at TRL 9 level is implemented and tested via another RPiZW node, a smartphone, and Twilio®'s as well as Gmail©'s APIs. Additionally, a non-web-based contingency device is developed using a Siemens® T35 GSM shield mounted on an Arduino® UNO™. The main objective with this implementation is to setup the foundations of an increasingly comprehensive intervention framework capable of reacting to emergency events, both with respect to the inhabitants of the built-environment and with this environment per se. The Twilio® implementation represents a cost-effective SMS service, while the T35 GSM setup represents a standard prepaid SMS service. A scenario may be entertained where the built-environment's WiFi service is unavailable for a period of time, yet the integrity of the WSAN's core (i.e., the local system) remains uncompromised as its constituents remain networked via ZigBee and BLE. In such a scenario, an emergency event may be reported via the T35 GSM setup, as it relies on standard cellular communication. Conversely, another scenario may also be entertained, where cellular services are unavailable due to lack of coverage. In this scenario, emergency events may be reported via Twilio®'s SMS service to any location worldwide. Both of these hypothetical scenarios presuppose that the recipient is capable of receiving cellular messages at the time of notification. However, this may not be the case. This kind of situation is the motivation behind email notifications. Although it cannot guarantee message reception, it adds yet another means for it. Unlike both SMS mechanisms, the email notification is free.

5.2.2.5   Machine Learning[10]

With respect to the first functionality, a *Machine Learning* (ML) subsystem is integrated in the proposed system-architecture in order to enable *Human Activity Recognition* (HAR) mechanisms (Liu Cheng et al. 2017). With respect to HAR, ML methods have typically used gyroscopic data collected via portable devices (e.g., smartphones, etc.) (Anguita et al. 2013; Ortiz 2015) or via sensor-fusion (Palumbo et al. 2016). The ML subsystem consists of two classification mechanisms developed based on polynomial programming of SVM and *k*-NN classifiers. These SVM and *k*-NN models are built on a dynamically clustered set of high-performance nodes in the localized WSAN.

---

[9]See (Liu Cheng et al. 2016) for a detailed discussion of this mechanism.

[10]See (Liu Cheng et al. 2017) for a detailed discussion of this mechanism.

Due to their evolving and resilient characters, ML classifiers have been implemented in a variety of applications built on WSANs (Alsheikh et al. 2014). HAR, as one such application, has successfully exploited classifiers in the last five years (see, for example, (Xiao and Lu 2015; Villa et al. 2012; Andreu and Angelov 2013). However, due to the cost-effective and low energy-consumption character typical of WSAN nodes, computational processing with respect to feature extraction has been considerably limited (Salomons et al. 2016). To overcome this limitation, the present implementation is capable of instantiating ad hoc clusters consisting of a variety of high-performance nodes. Furthermore, several clusters may be instantiated simultaneously in order to enable parallel high-performance information processing activities.

Another way to overcome this limitation is to avoid it altogether by outsourcing all high-performance information processing to cloud-based ML services.[11] But there are a number of limitations with this approach. The first, and perhaps the most salient, is the cost incurred by including proprietary services in any proposed intelligent built-environment solution. A second yet no less important limitation may be the impact to the solution's resilience. Should the built-environment lose access to the Internet, it would be incapable of generating classification models.

In the current state of D2RO, integration of both cloud-based as well as localized ML capabilities in order to ascertain robustness and resilience. Whenever possible, ML processes are locally and dynamically executed via ad hoc node-clustering. But should this prove impossible either due to failure or unavailability of proper resources, cloud-based ML services are used. More specifically, two ML mechanisms are integrated into the present system: (1) a localized ad hoc cluster system based on open-source and purpose-written *Python* scripts, and (2) a simulated cloud-based analytics service using MathWorks® MATLAB™. In both mechanisms SVM and *k*-NN classification models are generated.

In the localized mechanism, a script based on *pyOSC* is first written to receive OSC data from any device and application capable of broadcasting in described protocol. While all the WiFi-enabled nodes in the system's WSAN have the capacity to receive this data-streaming, only one of the nodes of the cluster instantiated to generate classification models stores it locally and streams it to a cloud-based data visualization service (i.e., Plotly™). Should the receiving node fail, another high-performance node will replace it automatically. Since the proposed solution uses a smartphone and three LBBs for data redundancy, resolution, and validation, the script in question proceeds to parse and to reduce the noise in the received multi-sensor data in order to generate a robust and unified dataset. At this point the dataset is processed through two ML scripts based on *scikit-learn* (Pedregosa et al. 2011; Buitinck et al. 2013), one for SVM and another for *k*-NN classification models.

---

[11]E.g., Google® CloudPlatform™, Amazon® Machine Learning™, Microsoft® Azure™, etc.

5.2.2.6 Object Recognition via OpenCV[12]

The object-recognition mechanism reaching TRL 9 is implemented with open-source BerryNet® (2017), which is built with a classification model (viz., Inception® ver. 3 (Szegedy et al. 2016) as well as a detection model (viz., TinyYOLO® Redmon and Farhadi 2017). The classification model uses *Convolutional Neural Networks* (CNNs), which are at the forefront of ML research (Szegedy et al. 2016). An advantage of BerryNet® is that it is a fully localized DL gateway implementable on a cluster of RPi3 s. On an individual RPi3, the *inference* process is slow, requiring a delay between object-recognition sessions. This situation is ameliorated by the dynamic clustering feature of the WSAN. Another benefit-*cum*-limitation is that BerryNet®'s classification and detection models are pretrained, which avoids the need to generate models locally.

The object-recognition mechanism in D2RO is intended to be deployed across a variety of cameras in the overall built-environment, and that instances of detection were to be cross-referenced to minimize false positives. In order to implement this setup, each RPi3 node in the WSAN is equipped with a low-cost Raspberry Pi Camera® V2.1, then BerryNet® is installed in every node and the *inference* mechanism tested individually. The next step is to enable the nodes to share their detection results, which could be done via WiFi. Nevertheless, in order to reduce energy-consumption for every object-detection cross-referencing instance, ZigBee is preferred. In order to enable ZigBee on BerryNet®'s *detection_server.py* and *classify_server.py* were modified and made compliant with *python-xbee* (2017).

5.3 Design-to-Robotic-Production and -Operation

The integration of D2RO with D2RP, as explored at TU Delft, relies on the notion of hybrid componentiality. This implies that components are cyber-physical and their design is informed by structural, functional, environmental, assembly and operation considerations (Mostafavi and Bier 2016). At the *micro*-scale, the material is conceived as a porous system, where the degree and distribution of porosity i.e. density are informed by functional, structural and environmental requirements, while taking into consideration both passive (structural strength, thermal insulation, etc.) and active (adaptive, reconfigurable, etc.) behaviours. At the *meso*-scale, the component is informed mainly by the assembly logic, while at the *macro* scale, the assembly is informed by architectural considerations.

D2RP&O has been explored in the project *Hybrid Assemblies* (see Fig. 5.8) implemented with students *Dessau Institute of Architecture* (DIA). The project focused on the development of architectural systems composed of heterogeneous components addressing various requirements from functional and formal to structural and climatic. While taking these requirements into account, the project focused on the

---

[12]See (Liu Cheng et al. 2017) for a detailed discussion of this mechanism.

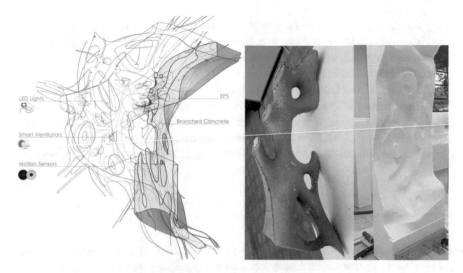

Fig. 5.8 Multi-layered D2RP&O integration logic (left) of fragment made of concrete (middle) that is cast in robotically produced EPS (right)

notion of embedded interactive or adaptive systems employed for climate control. The distributed, dynamic climate control has been conceived as consisting of intelligent networked climate control components, locally driven by people's preferences and changing environmental conditions. The challenge was to integrate the passive energy saving material architecture with the active climate control that is taking into account changes in the use of space and respective fluctuating needs based (not on average but) on real-time data.

The design was defined by optimization strategies involving spatial configuration, structural analysis, heating and cooling, lighting requirements, and the integration of ICT devices.[13] While, structural analysis is employed to map areas that are needed for structural support, lighting is determined based on 24/7 activities and their corresponding requirements. These inform the shape and the location of cavities for LED-based illumination. Then heating and cooling requirements are identified for the integration of intelligent ventilation systems as well as the required sensors for automated control.

The multi-layered hybrid components consisting of concrete, EPS, and smart devices follow *componentiality* and *hybridity* principles characteristic of D2RP&O. Layers are designed in direct response to a purpose or a function. For example, the concrete layer is formed following the stress lines and cavities in the EPS layer are designed according to ICT-integration requirements (Fig. 5.8). This approach embeds all cyber-physical requirements from the onset of the design process.

With respect to Systems Architecture, the detailed object-recognition mechanism adds another means for the system to become aware of the built-environment. In the

---

[13] See (Liu Cheng et al. 2016) for a detailed discussion of this mechanism.

setup discussed, the deployment scenario is construed as a single-occupant housing unit. But in scenarios with more occupants, the recognition of each individual may instantiate actuations and transformations in the built-environment specific to each individual's preferences.

The integration of D2RP with D2RO as explored at TU Delft is unprecedented in particular because of the focus on buildings. Installations such as Open Columns or the Hyozolic series, which reconfigure according to changing levels of CO_2 or movement of people, may be integrating computational design with additive manufacturing and smart reconfiguration but their application to buildings is still speculative.[14]

5.4   Conclusion

D2RP&O is unique in its aim to link design and production with smart operation of the built environment and advances applications in performance optimization, robotic manufacturing, and user-driven operation in architecture. Relying on human and non-human interaction in the design, production, and operation of buildings, D2RP&O is fundamentally changing the role of the architect. Architects design increasingly processes not objects, while users operate multiple time-based architectural configurations emerging from the same physical space that reconfigures in accordance to environmental and user specific needs. In this context, D2RP&O empowers architects to regain control over the design implementation into physically built environments and allows end-users to participate as co-creators in the adaptation i.e. customization of their environments over time.

Even if D2RP and D2RO have been developed as separate areas of research, their partial integration into a coherent D2RP&O chain has been implemented and tested in the *Hybrid Assemblies* project. This integration indicated that D2RP&O could significantly contribute to improving material-, energy-, and process-efficiency, as well as (structural, environmental, functional, etc.) performance of buildings.

In addition to developing a coherent D2RP&O chain, the challenge for the future is the integration of Human-Robot Interaction (HRI). For instance, by employing laser scanning to capture the current status of building process, an extended feedback-loop between the virtual and the physical environments is established. D2RP robots may then interact with humans, as for instance, human operators may teach robots to do certain tasks by guiding them with a tool or by hand, while dynamic safety systems are in place,[15] etc. Similarly, D2RO relies on HRI when sensor-actuators ensure that inhabitants can customize the use of the physically built space. Main consideration

---

[14]The two installations were developed as architecture inspired art projects (accessed from http://cast.b-ap.net/opencolumns/ and http://www.philipbeesleyarchitect.com/sculptures/0929_Hylozoic_Ground_Venice/).

[15]HRI is in detail described in the chapter titled "Human-Robot Collaboration and Sensor-Based Robots in Industrial Applications and Construction" of this volume.

is that production and operation of buildings will be in the future robotized and identifying which skills sets are better acquired and executed by humans while others by machines is key to developing interaction scenarios between humans and robots.

Acknowledgements This chapter is based on research that has been partially presented in conference and journal papers that authors refer to. It has, in particular, profited from the *Robotic Building* session at the Game Set Match#3 symposium organized at TU Delft, 2016 and builds up on the abstract presented at this session later on published in Spool's first issue on Cyber-physical Architecture. Presented research has been implemented with *Robotic Building* researchers from TU Delft and M.Sc. students from TU Delft and DIA. It has been supported and/or sponsored by 4TU, TU Delft, *Delft Robotic Institute*, 100% Research, DIA, ABB, and KUKA.

References

Abbate S, Avvenuti M, Bonatesta F, Cola G, Corsini P, Vecchio A (2012) A smartphone-based fall detection system. Pervasive Mobile Comput 8(6):883–899

Acampora G, Cook DJ, Rashidi P, Vasilakos AV (2013) A survey on ambient intelligence in healthcare. Proc IEEE 101(12):2470–2494

Alsheikh MA, Lin S, Niyato D, Tan H-P (2014) Machine learning in wireless sensor networks: algorithms, strategies, and applications. IEEE Commun Surv Tutor 16(4):1996–2018

Andò B, Baglio S, Lombardo CO, Marletta V (2014) An advanced tracking solution fully based on native sensing features of smartphone. In: 2014 IEEE Sensors Applications Symposium (SAS), pp 141–144

Andreu J, Angelov P (2013) An evolving machine learning method for human activity recognition systems. J Ambient Intell Hum Comput 4(2), 195–206. https://link-springer-com.ezproxy.librar y.ubc.ca/content/pdf/10.1007%2Fs12652-011-0068-9.pdf

Anguita D, Ghio A, Oneto L, Parra X, Reyes-Ortiz JL (2013) A public domain dataset for human activity recognition using smartphones. In: Proceedings of the 21th european symposium on artificial neural networks, computational intelligence and machine learning, ESANN 2013

ASHRAE® Standard 62.1-2013, 2013

ASHRAE® Standard 55-2013, 2013

Augusto JC, Novais P, Corchado JM, Analide C (eds) Ambient Intelligence and Future Trends: International Symposium on Ambient Intelligence (ISAmI 2010). Advances in Intelligent and Soft Computing, vol 72. Springer Verlag

Bier HH, Knight T (2014) Digitally-driven architecture (English). Footprint 4(1): 1–4

Bier HH, Knight T (2014b) Data-driven design to production and operation. Footprint 8(2):1–8

Bock T, Güttler J, Georgoulas C, Linner T (2015) The development of intra-house mobility, logistics and transfer solutions in PASSAge. JRM 27(1):108

Buitinck L, Louppe G et al (2013) API design for machine learning software: experiences from the scikit-learn project. http://arxiv.org/pdf/1309.0238

Comité Européen de Normalisation© (CEN) (2007) Standard EN 15251–2007: Indoor environmental input parameters for design and assessment of energy performance of buildings addressing indoor air quality, thermal environment, lighting and acoustics http://www.sysecol2.ethz.ch/Opt iControl/LiteratureOC/CEN_06_prEN_15251_FinalDraft.pdf. Accessed 16 July 2007

Cook P (1970) Experimental architecture. Universe Books, New York

Cook P (1972) Archigram. Studio Vista, London

Cook DJ, Augusto JC, Jakkula VR (2009) Ambient intelligence: Technologies, applications, and opportunities. Pervasive Mobile Comput 5(4):277–298

DT42©, Ltd, *BerryNet® (2017)* Deep learning gateway on Raspberry Pi. https://github.com/DT4 2/BerryNet. Accessed 21 Jun 2017

Eastman CM (1972) Adaptive conditional architecture. Institute of Physical Planning, School of Urban and Public Affairs, Carnegie-Mellon University, Pittsburgh

Esch J (2013) A survey on ambient intelligence in healthcare. Proc IEEE 101(12):2467–2469

European Association of Research and Technology Organisations (EARTO) (2015) The TRL Scale as a Research & Innovation Policy TOOL: EARTO Recommendations. http://www.earto.eu/file admin/content/03_Publications/The_TRL_Scale_as_a_R_I_Policy_Tool_-_EARTO_Recomm endations_-_Final.pdf. Accessed 07 Jan 2015

Fox M (2010) Catching up with the Past: A small contribution to a long history of interactive environments

Fox M, Kemp M (2009) Interactive architecture, 1st edn. Princeton Architectural Press, New York

Frazer J, Frazer J (1979) Letter To Cedric Price, Jan 1979, Box 5, Generator document folio, DR1995:0280:65 5/5, CPA

Georgoulas C, Raza A, Güttler J, Linner T, Bock T (2014) Home environment interaction via service robots and the leap motion controller. In: Proceedings of the 31st international symposium on automation and robotics in construction (ISARC 2014)

GitHub Inc.© (2017) alexa-avs-sample-app. https://github.com/alexa/alexa-avs-sample-app. Accessed 01 Apr 2017

Harrison BL, Consolvo S, Choudhury T (2010) Using multi-modal sensing for human activity modeling in the real world. In: Nakashima H, Aghajan HK, Augusto JC (eds) Handbook of ambient intelligence and smart environments. Springer, New York, pp 463–478

Jaskiewicz T (2013) Towards a methodology for complex adaptive interactive architecture

Jiménez-Fernández S, de Toledo P, del Pozo F (2013) Usability and interoperability in wireless sensor networks for patient telemonitoring in chronic disease management. IEEE Trans Bio-med Eng 60(12):3331–3339

Kolarevic B (2014) Outlook: Adaptive Architecture: Low-Tech, High-Tech, or Both?. In: Kretzer M, Hovestadt L (eds) Applied Virtuality Book Series, v.8, ALIVE: Advancements in adaptive architecture. Birkhäuser, Boston, pp 148–157

Latour B (2009) Reassembling the social: an introduction to actor-network-theory. Oxford University Press, Oxford

Linner T, Georgoulas C, Bock T (2012) Advanced building engineering: Deploying mechatronics and robotics in architecture. Gerontechnology 11(2):380

Liu Cheng A (2016) Towards embedding high-resolution intelligence into the built-environment. Archidoct 4(1), 29–40 http://www.enhsa.net/archidoct/Issues/ArchiDoct_vol4_iss1.pdf

Liu Cheng A, Bier HH (2016a) An extended ambient intelligence implementation for enhanced human-space interaction. In: Proceedings of the 33rd international symposium on automation and robotics in construction (ISARC 2016), pp 778–786

Liu Cheng A, Bier HH (2016b) Adaptive building-skin components as context-aware nodes in an extended cyber-physical network. In: Proceedings of the 3rd IEEE world forum on internet of things. IEEE, pp 257–262

Liu Cheng A, Bier HH, Mostafavi S (2017) Deep learning object-recognition in a Design-to-Robotic-Production and -Operation implementation. In: Proceedings of the 2nd IEEE Ecuador technical chapters meeting 2017, Guayaquil, Ecuador

Liu Cheng A, Georgoulas C, Bock T (2016) Fall detection and intervention based on wireless sensor network technologies. Autom Constr

Liu Cheng A, Bier HH, Latorre G, Kemper B, Fischer D (2017) A high-resolution intelligence implementation based on Design-to-Robotic-Production and -Operation strategies. In: Proceedings of the 34th international symposium on automation and robotics in construction (ISARC 2017)

Micucci D, Mobilio M, Napoletano P, Tisato F (2017) Falls as anomalies?: An experimental evaluation using smartphone accelerometer data. J Ambient Intell Humaniz Comput 8(1):87–99

Milgrom PR (1990) The economics of modern manufacturing: technology, strategy, and organization. Am Econ Rev 80(3):511–528

Mostafavi S, Bier H (2016) Materially informed design to robotic production: A Robotic 3D printing system for informed material deposition. In: Reinhardt D, Saunders R, Burry J (eds) Robotic fabrication in architecture, art and design 2016. Springer, Heidelberg, pp 339–349

Mostafavi S, Bier H, Bodea S, Anton AM (2015) Informed design to robotic production systems; developing robotic 3D printing system for informed material deposition. In: Proceedings of the 33rd eCAADe conference real time, vol. 2, Vienna (Austria), 16–18 Sept 2015 https://repository.tudelft.nl/islandora/object/uuid%3A02bf0811-41f0-4ac9-a270-af796fc77c7c/datastream/OBJ/download

Nakashima H, Aghajan HK, Augusto JC (eds) (2010) Handbook of ambient intelligence and smart environments. Springer, New York

Negroponte N (1969) Toward a theory of architecture machines. J Archit Educ (1947–1974) 23(2):9

Negroponte N (1975) The architecture machine. Comput Aided Des 7(3):190–195

n.io Innovation© (2017) LLC, python-xbee: Python tools for working with XBee radios. https://github.com/nioinnovation/python-xbee. Accessed 01 Jun 2017

Oosterhuis K (2012) Hyperbody: First decade of interactive architecture. Jap Sam Books, Heijningen

Oosterhuis K, Bier HH (2013) IA #5: Robotics in architecture. Jap Sam Books, Heijningen

Ortiz JLR (2015) Smartphone-based human activity recognition. Springer, Cham

Palumbo F, Gallicchio C, Pucci R, Micheli A (2016) Human activity recognition using multisensor data fusion based on Reservoir computing. AIS 8(2):87–107

Pask G (1975a) Conversation, cognition and learning: A cybernetic theory and methodology. Elsevier, Amsterdam

Pask G (1975b) The cybernetics of human learning and performance: A guide to theory and research. Hutchinson Educational, London

Pedregosa F, Varoquaux G et al (2011) Scikit-learn: Machine Learning in Python. http://arxiv.org/pdf/1201.0490

Pottmann H, Asperl A, Hofer M, Kilian A (2012) Architectural geometry. Bentley Institute Press, Exton

Price C (1978) Letter To John Frazer, Dec 1978, Box 4, Generator document folio, DR1995:0280:65 4/5, CPA

Redmon J, Farhadi A (2017) YOLO9000: Better, Faster, Stronger. http://arxiv.org/pdf/1612.08242

Salomons EL, Havinga PJM, van Leeuwen H (2016) Inferring human activity recognition with ambient sound on wireless sensor nodes. Sensors (Basel, Switzerland) 16(10)

Schnädelbach H (2010) Adaptive architecture-a conceptual framework. In: Proceedings of Media City

Seppänen O, Kurnitski J (2009) Moisture control and ventilation. World Health Organization

Steenson MW (2014) Architectures of information: Christopher Alexander, Cedric Price, and Nicholas Negroponte and MIT's Architecture Machine Group

Szegedy C, Vanhoucke V et al (2016) Rethinking the inception architecture for computer vision. http://arxiv.org/pdf/1512.00567

Villa AE (2012) Artificial neural networks and machine learning– ICANN 2012. In: 22nd International conference on artificial neural networks, Lausanne, Switzerland, 11–14 Sept 2012, Proceedings. Springer, New York

Xiao W, Lu Y (2015) Daily human physical activity recognition based on kernel discriminant analysis and extreme learning machine. Math Probl Eng 2015(1):1–8

Rajkumar R, Lee I, Sha L, Stankovic J (eds) Cyber-physical systems: the next computing revolution. ACM

Yang S-H (2014) Wireless sensor networks principles, design and applications. Springer, London

Zelkha E, Epstein B, Birrell S, Dodsworth C (1998) From devices to 'ambient intelligence': the transformation of consumer electronics (Conference Keynote). In: Digital living room conference

Chapter 6
Dispositions and Design Patterns for Architectural Robotics

Keith Evan Green

Abstract Embedding robotics in an architectural work lends the work a semblance of vitality: the capacity to move with and respond to things external to it. It is this capacity that defines Architectural Robotics (AR) and, potentially, forges more interactive, more intimate relationships between our physical surroundings and us. *Will human beings be prepared to inhabit this whirling space of physical bits and digital bytes?* Assuming an optimistic view, this chapter offers a response, drawing from art and art history, environmental design, literature, psychology, and evolutionary anthropology, to identify wide-ranging dispositions in humans for such "new places" of human-machine interaction. Additionally, this chapter offers a formal taxonomy of design patterns for AR. Research from the author's lab serve as design exemplars for future work by other design researchers.

6.1 Introduction

It is well recognized that human-computer interaction (HCI) today is no longer bound by computer displays ("one human–one computer") or by Weiser's vision of ubiquitous computing ("people connected by an invisible web"). Today, the horizons of human-computer interaction are defined, in part, by physical scale. At one end of the physical spectrum. where HCI approaches nothingness, computing resides not only around us but also *on* us and *in* us, embedded notably as a bionic second-skin forging a connection between our bodies and the external world (Someya 2013).

At the other end of the physical spectrum, computing is embedded in the very fabric of our everyday living environments, manifested as networked smart appliances (the Internet of Things [IoT]), physical and tangible computing (Tangible User Interface [TUI]), assistive, humanoid robots (Human-Robot Interaction [HRI]), and as shape-shifting furniture, rooms, building façades, and urban infrastructure (Architecture Robotics [AR]).

K. E. Green (✉)
Cornell University, Ithaca, NY, USA
e-mail: keg95@cornell.edu

© Springer International Publishing AG, part of Springer Nature 2018
H. Bier (ed.), *Robotic Building*, Springer Series in Adaptive Environments,
https://doi.org/10.1007/978-3-319-70866-9_6

An emerging subfield of Adaptive Environments (AE), Architecture Robotics (AR) is computing hardware made spatial and inhabitable. AR manifests itself as meticulously designed, inhabitable environments made interactive, adaptive, and at least partly intelligent. A key behavioral trait of AR is its capacity to respond and adjust to external, often dynamic inputs, whether these inputs are the needs and wants of human inhabitants, or changes in environmental or climactic conditions, or updated information supplied by the Internet. The response of AR to external input manifests itself primarily by changing shape, but this can be accompanied, also, by changes in color and sound. In the author's *Architectural Robotics Lab* at Cornell University, established in 2005 with collaborator Ian Walker at Clemson University, AR has assumed the form of: an Assistive Robotic Table ("ART") enabling, in particular, post-stroke patients (Threatt et al. 2014); an Animated Working Environment ("AWE") that re-conforms to support the working life of co-located, information Age workers working at once with digital and analog materials and tools (Houayek et al. 2014); and a LIT ROOM cultivating literacy in children by transforming the everyday space of the public library into the imaginary space of the book (Schafer et al. 2014).

Across the physical spectrum, recent triumphs in these new horizons of HCI nevertheless remind us of that old, unsettling adage: *Just because you can, doesn't mean you should*. The same sentiment has been attributed recently to assistive humanoid robots and Artificial Intelligence (AI), the latter which will form, likely, the glue that binds together the various scales of next horizon HCI artifacts to form cyber-physical (eco)systems [CPS] of smaller and larger, interactive and intelligent, computing artifacts. In this expanded CPS, the human users in HCI become inhabitants of a whirling world of physical bits, digital bytes, and their hybrids. This reconfigurable habitation begs the question (borrowing words from *Science* on the future of AI), "What will the world be like if [this kind of computing] comes to coexist with human kind?" (Antón et al. 2014). While the AI community addresses this question, some with fear, others with anticipation (IEEE 2008), the HCI research community appears more satisfied with reporting on research triumphs, neglecting meanwhile to consider the meta question, What is it about human beings and being human that compels these next horizons of HCI? Why make our designed environments reconfigure?

6.2 Dispositions for Architectural Robotics

Offered as an impetus for much needed self-reflection, this chapter is an effort to address this core question from a cautiously optimistic stance. While the philosophical (i.e. phenomenological) dimension has been adeptly addressed for HCI by Dourish (2001), the response here draws instead from art and art history, environmental design, literature, psychology, and evolutionary anthropology.

6.2.1   Drawing from Art and Art History

Imagine a collection of appliances (IoT) or a robotic workplace (IE) that intelligently reconfigures to support changes in the workflow, recognizing the need for a particular adaption or reconfiguration that will better support it. The design of such systems requires the design team to envision (theoretically) innumerable pathways to adaption and reconfiguration: to essentially recognize in one form still other forms. This is a very different way to think about form for designers where convention assumes that form is singular and stable. Art historian Henri Focillon thought other than conventionally, grappling with the notion that a single form is neither singular nor stable but rather has within it a multitude of forms. "Although form is our most strict definition of space," wrote Focillon, "it also suggests to us the existence of other forms" (Focillon 1989). We must "never think of forms, in their different states, as simply suspended in some remote, abstract zone; they mingle with life, whence they come; they translate into space certain movements of the mind" (Focillon 1989). As forms are conceived and engaged by their users, "each form," writes Focillon, "is in continual movement, deep within the maze of tests and trials" to which their users submit them (Focillon 1989). In art, perhaps the clearest statement of this reciprocity between the dynamism of form and human perception is found in Italian Futurism, the artistic movement of the early 20th century, evidenced by the words of the movement's founder, F. T. Marinetti: "A house in construction symbolizes our burning passion for the coming-into-being of things. Things already built and finished, bivouacs of cowardice and sleep, disgust us! We love the immense, mobile, and impassioned framework that we can consolidate, always differently, at every moment" (Marinetti 1991). The thinking of Focillon and Marinetti suggest to the designers of the next horizons of HCI that an artifact is not singular and isolated but an "open work," a kind of "hypertext," an artifact open to users' interpretations as imparted by memory and by the physical, virtual, and cultural contexts in which the artifact resides (Eco 1989).

6.2.2   Drawing from Environmental Design

With few exceptions, designing the built environment for movement, for *reconfigurability*, has been resisted by designers throughout history. Resistance to reconfigurability is motivated by the requirement of buildings to maintain continuity, to defy or at least to resist the impositions of nature and unfamiliar humankind. Curiously, today's homes and workplaces remain largely incapable of responding to changes occurring in their inhabitants as these inhabitants grow, grow old, and sometimes grow sick, and as groups of inhabitants grow and shrink in their numbers and exhibit varied and fluctuating needs and wants. Environmental design (mostly equated with architecture) has mostly ignored this flux endemic to life.

From the aesthetic, formal side, resistance to reconfigurability is motivated by the quest for a universal standard for measuring it: in designing its parts, and in organizing these parts to constitute the whole work. From architects Vitruvius to Le Corbusier—two millennia between them—the dimensional and proportional systems of buildings and other aspects of the built environment were modeled on an idealized and yet motionless human body: Vitruvian and Modular men. Maintaining the continuity bridging these two figures is the Renaissance ideal of a "timeless" and "beautiful" building in which "nothing may be added, taken away or altered" (Alberti 1988).

This "immobility" of architecture has its historical exceptions. It is not entirely novel for a piece of furniture or even a building interior to permit changes to its physical form to afford different functions supporting different human objectives or activities. These kinds of mechanical affordances or *action possibilities* date back centuries, for example, in the form of tatami mats and sliding shoji screens found in traditional Japanese houses. Most notably in architecture, the Rietveld Schröder House (1924, Utrecht), designed by cabinetmaker and architect Gerrit Thomas Rietveld, extended the concept of the sliding screen to permit the manual reconfiguration and repositioning of various components of the home's second story.

Carlo Mollino, a mid-twentieth century architect known for his own reconfigurable architectural contrivances, imaginatively characterized the manually reconfigurable house as "a jack-in-the-box, a play of easily changeable rooms and furnishings, a fickle scenography of embroidered furnishings and sliding, transforming rooms, separating and creating halls and lounges with the turn of the seasons, in states of animation, reflecting the ceremonies of 'domestic' happenings.... When importune, the furnishings truly disappear into the wall" (Mollino 1949). The "easily changeable rooms and furnishings" (Mollino 1949) that Mollino describes are alive with possibilities for reconfiguring them. What fascinated this Turinese architect was not so much the physical movement afforded by the sliding partitions and furnishings (their mechanics), but mostly how these architectural elements, in their flexibility, reflected things external to them: the passing of the seasons, the unfolding rituals of domestic life, our own inner selves. In the rooms and furnishings of his own design, Mollino invited inhabitants to tune the mechanical features of these strange places to reflect the conditions of their interior lives—to reflect themselves in the environments in which they live, *to make themselves more at home*. In the number of interior domiciles he designed for himself, the frenetic Mollino sought a sense of restfulness for himself, but recognized, in states of torment and elation, the difficulty of capturing this peace, even for the duration of the shutter movement of his Leica camera.

Despite the best efforts of environmental design, its works are no more static than the lives living within them. When we enter a building, we bring with us the dimension of time. No inhabitant will ever have precisely the same experience here, nor will any other inhabitant have precisely the same experience here as someone else inhabiting the same space. The human experience, framed by the physical environment, is never precisely the same at two points in time. A work that is reconfigurable is one that, at least in conventional architectural terms, is *unfinished*: room is made in the very design of such a place for the inhabitants to, in a word, play. Architectural works, like all works of art, are "quite literally 'unfinished,'" Umberto Eco contended: "the author seems to hand them on to the performer more or less like the components of a construction kit (Eco 1989). For Eco, as might be said for F. T. Marinetti, "the comprehension and interpretation of a form can be achieved only … by repossessing the form in movement and not in static contemplation" (Eco 1989). In the strange built environments described here, designers of IoT, IE, HRI and broadly CPS can discern compelling precedents for designing cyber-physical environments that actively grow and adapt with their users over time.

6.2.3 Drawing from Literature

The means of computing (including robotics) can be integrated into the physical fabric of things to forge a more interactive, more intimate relationship between the built environment and us. Embedding digital technologies in selected aspects of the built environment, from small appliances to the metropolis, renders these a semblance of vitality: the capacity to move with and respond to things external to them, whether these things are living (people and pets), or inanimate (physical property), or phenomena far less tangible (data streaming over the Internet, the detection of weather). In this very active way—of engaging the world, drawing inferences, and responding in kind—cyber-physical artifacts are, to a degree, a reflection of us: our needs, our aspirations as vital beings that "change shape".

As evidenced by its centrality in classical Greek mythology, "shape-shifting" has fascinated us for millennia. As Steven Levy asserts in *Artificial Life*, today's human-made, life-like artifacts are founded not only in the contemporary imagination, but equally so in the many "ancient legends and tales" devoted to the theme of "inanimate objects" infused "with the breath of life" (Levy 1993).

Following Levy and recognizing physical reconfigurability as a pathway, today, to a more intimate correspondence between our physical environments and ourselves, it is not such a stretch to learn from the myth of Proteus, the Greek god who, more than other shape-shifters in Greek mythology, was capable of transforming himself into countless different forms. This captivating capacity of Proteus to shape-shift led to his ultimate "transformation": into the familiar adjective, *protean*, which wonderfully captures a core behavior of the next horizons of HCI. Despite his advanced age and waning stamina, the Proteus of this poem of the eighth century B.C. has led an active and prolonged life under the same name but in different guises. Notably, Proteus is the name given to characters in Milton's *Paradise Lost* and in Shakespeare's *Henry VI* and *The Two Gentlemen of Verona*. Proteus is also the name given to historic warships (both USS *Proteus* and HMS *Proteus* of the Royal Navy), and to a novel, contemporary sailing vessel (the Proteus WAM-V, which features a reconfigurable hull that conforms to the surface geometry of water currents). Proteus is also the name given to, respectively, a medical syndrome popularly identified with "the Elephant Man," a bacterium having a remarkable ability to evade the host's immune system, and a family of flower having more than 1,400 varieties. Our fascination with shape-shifting is evidenced not only by this extended and variegated procession of forms under the name Proteus, but also by the contemporary usage of the word *protean*, defined by the *Oxford English Dictionary* as: *adopting or existing in various shapes, variable in form; able to do many different things;* and *versatile*. All of these definitions aptly describe the strivings of researchers engaged in developing the next horizons of HCI.

6.2.4 Drawing from Psychology

There remains one more Proteus that will prove useful in uncovering the promise of the next horizons of human-computer interaction: the Proteus of psychology. Both the Proteus of Heinrich Khunrath, the sixteenth century German physician-alchemist, and the Proteus of Swiss psychologist Carl Jung in the twentieth century personified the elusive unconscious. But for our purposes, the more useful Proteus is the one that names a contemporary, psychological profile considered by psychiatrist Robert Jay Lifton. In *The Protean Self: Human Resilience in an Age of Fragmentation*, Lifton characterizes this modern-day Proteus as "fluid and many sided" and "evolving from a sense of self [that is] appropriate to the restlessness and flux of our time" (Lifton 1994). This Proteus, a "willful eclectic," draws strength from the variety, disorderliness, and general acceleration of historical change and upheaval. As Lifton writes, "One's loss of a sense of place or location, of home—psychological, ethical, and sometime geographical as well—can initiate searches for new 'places' in which to exist and function (Lifton 1994). The protean pattern becomes a quest for 'relocation'" (Lifton 1994). According to Lifton, the protean self actively responds to life's challenges and opportunities—whether pedestrian (working life, family life) or grand-scaled (social, economic, political)—by seeking "new 'places'" best suited for improvement, advancement, or at least escape. For a research community invested in an adaptive environment, we discover in the Protean Self a human personality that is adaptive to and even drawn to flux and fluidity.

6.2.5 Drawing from Evolutionary Psychology

The protean way—to be fluid, resilient, and on the move—is not only a tactical, cognitive response to living today, but is, according to anthropology researchers Antón, Potts, and Aiello, *the outstanding trait distinguishing the human species*. The protean way is defined as "adaptive flexibility," the cornerstone of this new paradigm for human evolution, as published by these three researchers in the journal, *Science* (Antón et al. 2014). Antón, Potts, and Aiello find evidence for adaptive flexibility in all the "benchmarks" defining our species: "dietary, developmental, cognitive, and social" (Antón et al. 2014). Moreover, and critical to establishing the motivation for the next horizons of HCI, adaptive flexibility in the human species arose in response to "environmental instability" (Antón et al. 2014). As argued by Antón, Potts, and Aiello, the human species did not evolve in "a stable or progressively arid savanna" as suggested in earlier paradigms of evolution, but rather "in the face of a dynamic and fluctuating environment" composed of "diverse temporal and spatial scales" (Antón et al. 2014). What distinguishes humans from other mammals is our adaptive flexibility, the capacity to "buffer and adjust to environmental dynamics" (Antón et al. 2014). The significance for our research community is clear: the human species is super-adaptive to "diverse spatial scales" and "environmental dynamics" (Antón et al. 2014). This new paradigm for evolution, along with Lifton's concept of the Protean self, suggest that we are prepared for, and can in all probability make use of, controlled reconfigurations and adaptions of cyber-physical ecosystems under those life circumstances that warrant their application.

6.3   Design Patterns for Architectural Robotics

Cutting across the diverse perspectives briefly surveyed here, from art and art history, to environmental design, to Greek mythology, to psychology, and to human evolution, is a recognition of the vibrant exchange between the dynamic world in which we live and the intimate and social nature of our being. Central to what it means to be human is to be fluid, resilient, and on the move. The next horizons of human-computer interaction, borrowing Lifton's words, have the potential to cultivate "many and new places" for individuals and groups of individuals facing wide-ranging challenges and opportunities.

What are the opportunities for Architectural Robotics? Overall, the human population is growing older, greener, more mobile, and more digital, which compels AR applications for the home, the workplace, the school, and the clinic. Moreover, where space is scarce and expensive, AR may be configured or configure itself to accommodate more than one of these applications within the same physical frame.

For the research community focused in AR and more broadly, adaptive environments, there are at least a number of ways to arrive at such "many and new places" within the same physical frame: by selecting a new place among programmed places to match life needs and opportunities (a mechatronic approach); by fine-tuning and then saving patterns of adaption and configuration to create new places (an interactive approach); and by allowing the cyber-physical environment to anticipate needs and wants, reconfiguring itself a new place for us (an intelligent approach).

Christopher Alexander's *A Pattern Language* (1977) offers AR a foundation in the author's suggestion for "compressing" two or more use patterns into a single, physical space. So while *A Pattern Language* is a catalogue of design patterns, each of which, as Alexander explained it, "represents our current best guess as to what arrangement of the physical environment will work to solve…a problem which occurs over and over again in our environment," Alexander hints at something beyond the singular pattern: "compressing" two or more patterns into a single space. As elaborated by

Alexander, "this compression of patterns illuminates each of the patterns, sheds light on its meaning; and also illuminates our lives, as we understand a little more about the connections of our inner needs."

To illustrate this compression of patterns, Alexander envisioned all of the functions of a typical house occurring in the space of a single room, resulting in a building that, in practical terms, exhibits an "economy of space" that is potentially "cheaper" to realize. A relatively recent demonstration of Alexander's "compressed" house is the "Domestic Transformer," a 330-square-foot, single-room home in Hong Kong of sliding walls and hinged panels, manually reconfigured by its owner-architect, Gary Chang, to fashion any one of twenty-four different living patterns. But while Chang's home is compelling and informative for AR, it is not adaptive—its reconfiguration is not accomplished by way of computation or electronics, but is performed manually by its human inhabitants. And for Alexander, beyond flexibility, compactness, and potential cost savings (as relevant to this Hong Kong apartment), a "compressed" home should be fundamentally "poetic," offering in its compacted, patterned layers a "denser" meaning for its inhabitants. Overall, *A Pattern Language* lends to the emerging field of Adaptive Environments the conceptual framework of a carefully conceived, physical environment affording intimate and evolving relationships across people, things, and physical space.

6.4 A Taxonomy of Design Patterns for AR and Examples of Each

This chapter has so far offered motivations for AR and has provided a conceptual foundation of AR informed by Alexander's "compressed" home. The three design patterns presented in this section—the reconfigurable, the distributed, and the trans-figurable—give form, figuratively and more literally, to AR. The characterization of each of the three design patterns is followed by a design exemplar that was developed by my lab over the past dozen years.

6.4.1 The Reconfigurable Environment ("A Room of Many Rooms")

Essentially Alexander's concept of "compressed patterns," where all the functions of a typical house occur in the space of a single room, the Reconfigurable Environment (Fig. 6.1) is a malleable, adaptive environment specifically dependent on moving physical mass to arrive at its shape-shifting, functional states supporting commonplace human activities. The Reconfigurable Environment is characterized by a continuous, compliant surface that renders the rooms relatively soft compared to the conventional, rectangular room. In a Reconfigurable Environment, (most) every-

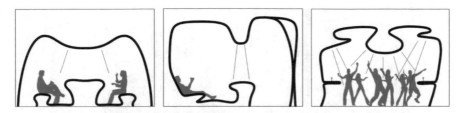

Fig. 6.1 The *Reconfigurable Environment*, a design pattern for AR in which one continuous envelope reconfigures to support widely different human activities such as eat/speak, lounge and play within the space of one room

thing across its continuous surface is capable of physically reconfiguring to create various configurations that evoke, yet transcend walls, ceiling, floor and furnishings accommodating the activities of its inhabitants.

As with all three design patterns characterized in this section, the Reconfigurable Environment is subject to "shared control" between the user and itself: the environment is not strictly intelligent, nor do users commandeer it. There is, instead, a degree of control, along a sliding scale, so that users are best served by this kind of human-machine rapport. This means that while the Reconfigurable Environment knows something about its component parts, it knows only a little something about its inhabitants, and reconfigures itself fittingly.

Figure 6.1 is meant to capture the essential character of the Reconfigurable Environment typology. Each of the three cells in the figure represent a brief period in the everyday life of the inhabitants living within this single volume, as its continuous surface morphs to allow them to play, lounge, eat and speak. The remarkable feature of the Reconfigurable Environment typology is that a single, physical space makes room for many places in support of and extending human activity.

6.4.1.1 An Example: *Animated Work Environment*

The Animated Work Environment or "AWE" (Fig. 6.2) is an interactive and user-programmable workplace environment that literally shapes the working life of multiple users co-located in a single, physical space, working separately and collaborating together. AWE supports and augments a 21st century workflow, where knowledge workers are engaged in complex tasks requiring non-trivial combinations of digital and physical artifacts, materials and tools, and peer-to-peer collaboration.

While more of us are caught up in cyberspace, we nevertheless continue to find utility and value in working with and generating physical things (Malone 1983). We also maintain the need for and desire for close collaborations with others, engaging together in complex work and leisure activities co-located in a single physical space. Indeed, ethnographic studies have shown consistently that people performing complex, creative tasks vigorously resist the "paperless office," preferring paper over computer tools for its ease with annotating, reconfiguring, organizing information

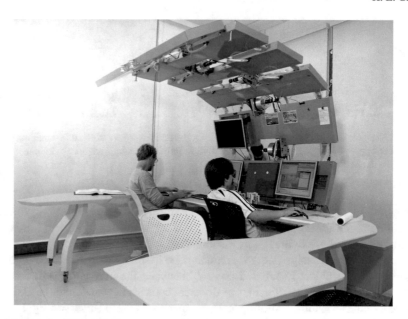

Fig. 6.2 *Animated Work Environment*

spatially, and shifting between storage, active use, and imminent use (Sellen and Harper 2002).

AWE has two key physical elements: a user-programmable, robotic display wall equipped with an array of embedded sensors and digital peripherals, and a programmable work surface (comprised of three table-like components) which is itself reconfigurable. In designing the Animated Work Environment (AWE), we sought to respond to these concerns by designing a workstation to meet two key goals: (1) mixed-media use—allowing users to use a range of digital and analog displays such as monitors, paper, whiteboards, and corkboards; and (2) user-programmability (reconfigurability)—allowing users to flexibly rearrange digital and analog display areas to meet changing task demands (Houayek et al. 2014).

AWE is viewed as part of a growing tendency within computing research that is concerned with various cross-cutting issues related to working life, including the use of multiple displays, managing mixed media, viewing healthcare information, and, more broadly, practices defined as Computer-Supported Collaborative Work (CSCW) (Baecker 1993). In particular, AWE builds on contemporary workspace design and on prior developments of interactive workplaces with embedded digital technology, such as the *Interactive Workspaces Project* (Johanson et al. 2002) and *Roomware* (Streitz et al. 2001). Precedents from workspace design, however compelling, focus not on automated or physically reprogrammable spaces but mostly on beautifully designed furniture (without embedded electronics) that support conventional ways of working. The informative precedents from HCI and interaction design, meanwhile, are mostly defined by collections of computer displays, smart boards, and novel peripherals that create electronic meeting rooms. AWE sits between these two tendencies, at the interface between computer technology, architectural design, and automation where the physical environment (including display surfaces for paper) is also subject to physical manipulation.

An exemplar of the Reconfigurable Environment typology that makes "room for many rooms," AWE affords wide-ranging workplace activities—individual work, co-authored work, prototyping, conferencing, presentations, serious gaming—to occur in the physical space of the common office cubicle or the smallest private office.

6.4.2 The Distributed Environment (in Which "Furnishings Come to Life")

The Distributed Environment (Fig. 6.3) that, seemingly, comes to life is comprised of individual, furniture-like components forming a suite of physically distributed furnishings. Typologically, the components of the Distributed Environment may look like familiar furnishings (say, a chair, desk, or sofa), or alternatively may look like two or more familiar furnishings joined together or hybridized as a single unit. In any case, the single unit of furniture physically reconfigures by the means of embedded robotics in response to human interactions with it, and/or by some input from the Internet or from its physical surroundings. Embedded robotics expands the affordances of such interactive furniture, as compared with the familiar affordances of static, single-purpose, conventional furniture.

As the contents of the Distributed Environment are identifiable as discreet, physical units, you can very much recognize the room—the physical envelope or container—for what it is: most commonly, rectangular in plan and of a familiar height. Figure 6.3 is meant to capture the essential character of the Distributed Environment typology, in which the three cells in the figure, furnished with physically distributed, morphing furniture, represent three everyday instances during which inhabitants play, lounge, eat and speak. While the individual furnishings that constitute the Distributed Environment are physically discrete, each one is networked with the others, and consequently knows something about the larger suite of furnishings, acting fittingly. In

Fig. 6.3 The *Distributed Environment*, a design pattern for AR in which physically distinct components reconfigure to support widely different human activities such as eat/speak, lounge and play

this respect, all three typologies considered in this chapter can be described as "distributed environments" in the sense that computing is distributed across them; but only the Distributed Environment is characterized by discrete, physical components distributed spatially. One notably exception is when the furnishings of the Distributed Environment are characterized by modular robotics, in which case (some or all) of the furnishing units are capable of physically connecting to one another. Yet even in the case of modular robotic furnishings, the individual physical units (the physical modules) tend to be visually discernable to users as discrete entities.

6.4.2.1   An Example: *Home+*

An exemplar of the Distributed Environment typology, *home+* is defined by a collection of networked, cyber-physical devices, each having different functionalities tuned for different purposes to support a common human need or to exploit an opportunity to improve the lives of its inhabitant(s). home+ is comprised, on the surface, of the commonplace furniture of an ordinary, contemporary house. The transformation to home+ is achieved, borrowing the words of William Mitchell, by way of "geographically distributed assemblages of diverse, highly, specialized, intercommunicating artifacts" that render the physical environment a "robot for living in" (Mitchell 1999). In similar terms, former Wired editor Kevin Kelly imagines a future artificial "ecology" of intelligent "rooms stuffed with co-evolutionary furniture" and a "mob of tiny smart objects," all having an "awareness of each other, of themselves, and of me" (Kelly 1994).

In recent years we have developed a number of home+ prototypes, including a robotic chair, robotic wall partitions, robotic tables, and a robotic lamp. We conceptualize the "animated" furniture of home+ and their users as "cohabitants" sharing a home together. home+ strives to empower people to remain in their homes for as long as possible, even as their physical capabilities alter over time, and, in more grave circumstances, to afford people some semblance of feeling at home as users move between their dwellings outfitted with home+ and assisted-care facilities equipped the same way. Our current and most advanced home+ prototypes are a robotic pair—a mobile, robot-cube and a continuum-robotic lamp that we call *h+ cube* and *h+ lamp* (Fig. 6.4), *h+ cube* and *h+ lamp* are designed to work as complements within the volume of the typical room of the home or workplace: *h+ cube* engages in tasks within a spatial volume bounded by a room's floor upwards to approximately hip or table height, while *h+ lamp* engages in tasks from hip or table height upwards to near the ceiling of a typical room.

Practically, *h+ cube* is a mobile robot that lifts objects from directly beneath it by way of linear actuators at each of its four legs working in concert with a jamming gripper, located at the core of the cubic volume and extending from a fifth, shorter linear actuator designed to rotate 90° and 180°. The ensemble is capable of transporting the retrieved object to a shelf, to a human co-habitant, or to the terminus of *h+ lamp*. Tugged about the interior living or work space by *h+ cube* and deposited by *h+ cube* where needed, *h+ lamp* has a mast comprised of a linear actuator closest to

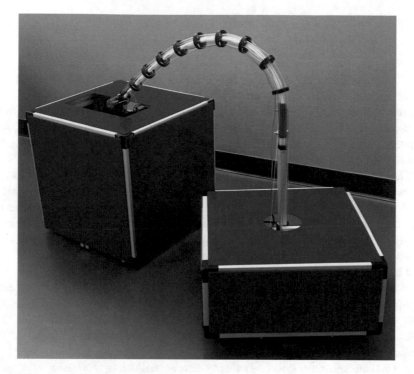

Fig. 6.4 *h+cube* (left) and *h+lamp* (right)

the floor extending continuously to a tendon-actuated, flexible continuum arm that is terminated by a two-finger gripper, a jamming gripper, and a bright light as means for illuminating task spaces. The mast ensemble can reach from below the top of the *h+cube* to the highest shelf in most homes or offices to transport objects across living and work spaces. *h+cube* can itself transport larger bins filled with objects (e.g. laundry or groceries) atop it. *h+cube* and *h+lamp* work together with their human co-inhabitants to perform ten routine tasks enumerated in the "CS-PFP10" scale widely used by healthcare providers to define the capacity for independent living.

Comprised of multiple, commonplace domestic artifacts, the Distributed Environment manifests something of the behavior of swarms, and also subscribes to the notion of the a-life community that, "in living systems, the whole is more than the sum of its parts" (Levy 1993). The remarkable feature of the Distributed Environment typology is that, seemingly, the everyday furnishings found at home, at work, at school and at still other common places "come to life," forming a "living" room.

6.4.3 The Transfigurable Environment (A "Portal to Elsewhere")

The Transfigurable Environment is that *other place*. And that *other place* may be the dunes on a beach with strong sea breezes, a bit of cool mist, and then, some sunbreaks through fast moving clouds (Fig. 6.5). In this way, the Transfigurable Environment is very much what Nicholas Negroponte defined as a "simulated environment" where "one can," for example, "imagine a living room that can simulate beaches and mountains" (Negroponte 1975).

The evocative forms of the Transfigurable Environment emerge from the walls, floors and/or ceilings, as do the functional forms, the furniture-like elements of the Reconfigurable Environment. However, as the Transfigurable Environment is not providing useful furnishings but rather evocations of someplace far outside the familiar, the conventional rectangular room it may occupy, along with any semblance of furnishings it may contain, dissolve away. The room is more womb or bladder than bedroom.

In the most optimistic and awe-inspiring way, the Transfigurable Environment is meant to follow from the well known Latin saying, *Vulgus vult decipi, ergo decipiatur* (People want to be deceived, so deceive them); it invites its inhabitants to be transported from the familiar to the unfamiliar, from reality to illusion. But the distinction between *unfamiliar* and *familiar*, *reality* and *illusion*, is not so clear here. As Albert Einstein understood, "Reality is merely an illusion, albeit a very persistent one."

Unmistakably, the Transfigurable Environment isn't charged with providing us the comforts of home. Quite the contrary: the Transfigurable Environment propels us from the comfort zone to unchartered territories, all in one space; because finding ourselves elsewhere is another (strange) form of coming home, as Max does within the confines of his bedroom, inside Sendak's book, after an adventure on the high seas. *The remarkable feature of the Reconfigurable Environment typology is that a single physical space becomes "a portal to elsewhere."*

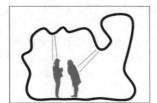

Fig. 6.5 The *Transfigurable Environment*, a design pattern for AR in which one continuous envelope reconfigures to transport inhabitants somewhere else

6.4.3.1   An Example: *LIT ROOM*

Today's digital technologies combined with meticulous interaction design provide a means for transporting us to the world of books and to larger realm of the imagination. The "simulated environments" envisioned by so many cultural fields for so long, and initially conceived by Negroponte and the Soft Architecture Machine Group (Negroponte 1975) are more accessible than ever.

Inspired by Negroponte's vision of a "living room that can simulate beaches and the mountains" (Negroponte 1975), the LIT ROOM exemplifies the Transfigurable Environment typology of Adaptive Environments, a "portal to elsewhere," embedded in the physical space of the library. As a robotic room, the LIT ROOM (Fig. 6.6) is transformed by words read from a picture book so that the everyday space of the library "merges" with the imaginary space of the book. The book is a room; the room is a book. A mixed-technology system for enhancing picture-book reading, the LIT ROOM combines the printed page with a multimodal, programmable experience evoking the book being read. The room-filled audio-visual-spatial effects of the LIT ROOM contextualize language and provide feedback to the participants. The LIT ROOM aims to scaffold critical literacy skills such as vocabulary acquisition, reading comprehension, and print motivation by creating a fun, interactive experience for children.

In the plainest terms, the LIT ROOM is a suite of four panels having the dimension of a very wide door or a very tall window (3-1/2-feet wide × 1-foot deep × 7-1/2-feet high), supported by customized Bosch Rexroth aluminum framing that forms a room-scaled, rectangular volume, 12-feet wide × 12-feet deep × 8-feet high. Embedded in each of the four panels are shape-changing, continuum-robot surfaces employing tendons for their reconfigurability. Additionally, the four panels are embedded with LED lighting, audio speakers, and associated electronics. At the center of the installation is a small, low-lying, circular table housing a conventional tablet computer. This table with embedded tablet operates as a lazy-Susan, rotating about its center axis to provide access to it for all LIT ROOM participants gathered around it. All of what comprises the LIT ROOM, sits within a conventional library interior, or within some other appropriate, physical environment.

More conceptually, the LIT ROOM is inspired partly by Ishii, Antle, and Reggio Emilia. Following from Ishii's sentiment (Ishii and Ullmer 1997) that HCI design suffers from a "lack of diversity of input/output media" and "a bias towards graphical output at the expense of input from the real world," the LIT ROOM conflates physical and cyber space while providing a rich media palette for play. Following from Antle's concept (Antle 2009) of created meaning "through restructuring the spatial configuration of elements in the environment," the robotics-embedded LIT ROOM environment is physically reconfigured by children, rendering it "co-adaptive": in course, both the environment and its inhabitants are transformed. This co-adaption in the LIT ROOM effectively reworks the Reggio philosophy (Rinaldi 2005), making it reflexive: so while "the environment is a teacher" to the child, the child is given ample opportunity to teach or cultivate the form the environment.

Fig. 6.6 *LIT ROOM*

In conceiving the LIT ROOM, my research team and I hypothesized that a Trans-figurable Environment, one that transports you elsewhere, cultivates learning. For this, we found validation, on a fundamental level again, in the findings previously considered: that human beings have evolved as protean super-adaptors who seek, acclimate to, and thrive in wide-ranging physical environments. The Muscle Body by Hyberbody at TU Delft exemplifies the Transfigurable Environment. Here, a playful, flexible, Lycra envelope reconfigures by McKibben actuators in response to the activities of the "players" inhabiting it. In the LIT ROOM, the Transfigurable Environment is tuned specifically as a learning environment, where there is ample evidence that changing the environment in which people learn fosters learning by forming new associations in their brains. As Benedict Carey, the author of *How We Learn* (Carey 2014), conveys, "The brain wants variation. It wants to move."

6.5 Many and New Places for Adaptive Beings

Architectural Robotics promises to provide inhabitants the means for creating a careful balance between stability and flexibility in a given moment. At their best, these "many and new places" will afford inhabitants the capacity, borrowing Lifton's words again, to "modify the self to include connections virtually anywhere while clinging to a measure of coherence" (Lifton 1994).

What this chapter strives to offer is the recognition that we and the cyber-physical (eco)systems on the near horizon are well matched: diverse, dynamic, adaptive and sometimes blurred. Manifested as health-care facilities, classrooms, workspaces, assisted-care homes, and potentially as mass public transit and road systems (traversed by autonomous cars), the next horizons of HCI will collapse further the boundaries that distinguish us from our surroundings when the conditions suggest (we hope) the greatest benefit to the individuals and the groups inhabiting them.

Obviously, the short space of a book chapter is woefully inadequate to elaborate, from six disciplinary perspectives, the motivations for the next horizons of HCI. The intent here, more so, is to offer the adaptive environments research community the impetus to reflect—to assume the "1000-mile view" that permits us *to see* (what we're doing), and *to recognize where we are*. From this vantage, Ivan Illich saw in new technology "tools of conviviality" fostering "self-realization" and "play" (Illich 2009). Buckminster Fuller saw a "spaceship earth" that lacked an operating manual that he could write, informed by "long-range, anticipatory, design science" characterized by "comprehensive thinking" (Fuller 2014). This author sees, with John Cage as guide, "gardens of technology" (Cage 1980) where every "inanimate object has a spirit" (Cage 1990). *What do you see?*

References

Alberti LB (1988) On the art of building in ten books (trans: Rykwert J, Leach N, Tavernor R). MIT Press, Cambridge

Antle AN (2009) Embodied child computer interaction: why embodiment matters. Interact 16(2):27–30

Antón SC, Potts R, Aiello LC (2014) Evolution of early homo: an integrated biological perspective. Science 345(6192):1236828

Baecker RM (1993) Readings in groupware and computer supported cooperative work. Morgan Kaufmann, San Mateo

Cage J (1980) Mesostic for Elfriede Fischinger. Center for Visual Music, Elfriede Fischinger Collection, Los Angeles

Cage J (1990) I–VI: the Charles Eliot Norton lectures. Harvard University Press, Cambridge

Carey B (2014) How we learn: the surprising truth about when, where, and why it happens. New York: Random House

Dourish P (2001) Where the action is: the foundations of embodied interaction. MIT Press, Cambridge

Eco U (1989) The open work. Harvard University Press, Cambridge. Trans. A. Cancogni

Focillon H (1989) The life of forms in art. Zone Books, New York

Fuller B (2014) [1969] Operating manual for spaceship earth. Lars Müller, Zurich

Houayek H, Green KE, Gugerty L, Walker ID, Witte J (2014) AWE: an animated work environment for working with physical and digital tools and artifacts. Pers Ubiquitous Comput 18:1227–1241

IEEE (2008) The singularity: special report. IEEE Spectrum (June). http://spectrum.ieee.org/stati c/singularity

Illich I (2009) [1973] Tools for conviviality. Marion Boyars, London

Ishii H, Ullmer B (1997) Tangible bits: towards seamless interfaces between people, bits and atoms. In: Proceedings of CHI 1997, the ACM Conference on Human Factors in Computing Systems. New York: ACM, pp 234–241

Johanson B, Fox A, Winograd T (2002) The interactive workspaces project: experiences with ubiquitous computing rooms. IEEE Pervasive Comput 1(2):67–74

Kelly K (1994) Out of control: the rise of neo-biological civilization. Addison-Wesley, Reading

Levy S (1993) Artificial life: a report from the frontier where computers meet biology. Vintage Books, New York

Lifton RJ (1994) The protean self: human resilience in an age of fragmentation. Basic Books, New York

Malone TW (1983) How do people organize their desks? Implications for the design of office information systems. ACM Trans Office Inf Syst 1(1):99–112

Marinetti FT (1991) 'The Birth of a Futurist Aesthetic' from war, the world's only hygiene [1911–1915]. In: Let's murder the moonshine: selected writings (trans: Flint RW). Sun and Moon, Los Angeles

Mitchell WJ (1999) E-Topia: urban life, Jim—but not as we know it. MIT Press, Cambridge

Mollino C (1949) Il Messaggio Dalla Camera Oscura ["The Message from the Dark Room"]. Turin: Chiantore, p 76

Negroponte N (1975) Soft architecture machines. MIT Press, Cambridge

Rinaldi C (2005) In dialogue with Reggio Emilia. London: Routledge

Schafer GJ, Green KE, Walker ID, Fullerton SK, Lewis, E (2014) An interactive, cyber-physical read-aloud environment: results and lessons from an evaluation activity with children and their teachers. In: Proceedings of DIS 2014, the ACM conference on the design of interactive systems. ACM, New York, pp 865–874

Sellen AJ, Harper Richard H R (2002) The myth of the paperless office. MIT Press, Cambridge

Someya T (2013) Building bionic skin. IEEE Spectr 50(9):50–56

Streitz NA, Tandler P, Müller-Tomfelde C, Konomi S (2001) Roomware: toward the next generation of human-computer interaction based on an integrated design of real and virtual worlds. In: Carroll J (ed) Human-computer interaction in the new millennium. Addison-Wesley, Boston, pp 553–578

Threatt AL, Merino J, Green KE, Walker ID, Brooks JO, Healy S (2014) An assistive robotic table for older and post-stroke adults: results from participatory design and evaluation activities with clinical staff. In: Proceedings of CHI 2014, the ACM conference on human factors in computing systems. ACM, Toronto, pp 673–682

Chapter 7
Movement-Based Co-creation of Adaptive Architecture

Holger Schnädelbach⬤ and Hendro Arieyanto

Abstract Research in Ubiquitous Computing, Human Computer Interaction and Adaptive Architecture combine in the research of movement-based interaction with our environments. Despite movement capture technologies becoming commonplace, the design and the consequences for architecture of such interactions require further research. This chapter combines previous research in this space with the development and evaluation of the MOVE research platform that allows the investigation of movement-based interactions in Adaptive Architecture. Using a Kinect motion sensor, MOVE tracks selected body movements of a person and allows the flexible mapping of those movements to the movement of prototype components. In this way, a person inside MOVE can immediately explore the creation of architectural form around them as they are created through the body. A sensitizing study with martial arts practitioners highlighted the potential use of MOVE as a training device, and it provided further insights into the approach and the specific implementation of the prototype. We discuss how the feedback loop between person and environment shapes and limits interaction, and how the selectiveness of this 'mirror' becomes useful in practice and training. We draw on previous work to describe movement-based, architectural co-creation enabled by MOVE: (1) Designers of movement-based interaction embedded in Adaptive Architecture need to draw on and design around the correspondences between person and environment. (2) Inhabiting the created feedback loops result in an on-going form creation process that is egocentric as well as performative and embodied as well as without contact.

H. Schnädelbach (✉)
Mixed Reality Lab, School of Computer Science, University of Nottingham, Nottingham, UK
e-mail: holger.schnadelbach@nottingham.ac.uk

H. Arieyanto
Casajardin Residence Software Maintenance, Jakarta Raya, Indonesia

© Springer International Publishing AG, part of Springer Nature 2018
H. Bier (ed.), *Robotic Building*, Springer Series in Adaptive Environments,
https://doi.org/10.1007/978-3-319-70866-9_7

7.1 Introduction

The concerns of Architecture, Ubiquitous Computing and Human Computer Interaction research have begun to overlap. Historically, this has been enabled by computing moving from the desktop into the environment via the emergence of ubiquitous and pervasive computing. Technically, this enabled sensors, actuators, processing and the interfaces to these to be embedded into the fabric of our surroundings, originally designed to function invisibly and to free us from performing mundane tasks (Weiser 1991). Because of these developments, researchers now address Human Computer Interaction in a much broader, considering the environment and artefacts as in tangible computing (Ishii and Ullmer 1997). Rogers frames this development as moving from users to context, employing multi-method study approaches, integrating knowledge from multiple disciplines to develop engaging user experiences that are evaluated through a value-focused lens (Rogers 2009). As a consequence, HCI research, drawing on Ubiquitous Computing technologies, is now frequently occupied with understanding interaction in the environment, a concern that Architecture traditionally holds.

As Coyne has recently re-stated, Architecture completely un-augmented by computation is already highly interactive (Coyne 2016). This is not to say that architects have not so far considered the inclusion of sensors, actuators and processing in their buildings. Quite the opposite is the case, and office buildings for example have been equipped for a long time so that their indoor climate can be tightly controlled, this history having been traced by Banham (1984). Today, eco-homes are commonly fitted with computer-controlled equipment with the aim to support people in reducing their carbon footprint and in creating a healthy and comfortable living environment. In addition, the rapid growth of the Internet of Things, results in ordinary homes to be augmented with digital technologies on an even larger scale. Beyond these more common examples, architectural research today includes interaction enabled through computation as an elementary part of its design palette. This has lead to a set of related publications addressing interactive (Fox and Kemp 2009), responsive (Bullivant 2005) and robotic (Bier 2014) architecture, which can be summarised as constituting the field of Adaptive Architecture (Schnädelbach 2010). Some of the key properties of such architecture are briefly described in what follows.

Sensing embedded in the environment but also body-worn (and communicating with the infrastructure embedded in the environment), provides information about people's location, movement, physiological, mental and psychological states, and their identity. *Actuation* in the architectural environment can be concerned with the light and sound infrastructure, environmental controls, data flow and media displays, resource supply and architectural components and elements, including their movement. In principle, anything that can be sensed about people can be linked to actuations in the environment. When such actuations are made, a *feedback loop* emerges between people's behaviour and the behaviour of the environment. Such feedback loops have been demonstrated in eco home research (Hong et al. 2016) as well as bespoke lab experimentation (Schnädelbach et al. 2012).

One focus of Adaptive Architecture research is the interaction between human movement and movement present in the environment. It is timely that this is considered in more detail, as architecture includes kinetic elements in more cases and sensor systems to capture people's movement are becoming more capable and widespread. As the number of prototypes that link human movement and architectural movement increase, the likelihood of such designs emerging in everyday buildings increases. In this context, it is essential that Architecture and Interaction research develop a better understanding of the opportunities and constraints that this brings and this chapter contributes to the development of this knowledge.

7.2   Background

The following briefly reviews existing work in mapping human and architectural movement.

7.2.1   Movement in Architecture

Everybody knows—and especially architects, of course—that a building is not a static object but a moving project, and that even once it is has been built, it ages, it is transformed by its users, modified by all of what happens inside and outside, and that it will pass or be renovated, adulterated and transformed beyond recognition. (Latour and Yaneva 2008).

While Latour's essay frames this as a representational problem, i.e. he is mainly concerned with how we might capture the fact that buildings are not static, works by Duffy (1990) and Brand (1994), have explored the more practical sides of which aspects of the built environment change over time, with those of larger scale and shared moving less rapidly than those smaller items that belong to individuals. In recognition of the non-static nature of buildings, Habraken has encapsulated this in his 'Supports' strategy, combining mass-production of supporting frames with individually adaptable dwelling units to enable adaptation over time (Habraken 1972).

There is also a class of buildings that are specifically designed to be mobile and the history of portable architecture has been succinctly captured by Kronenburg (2002). More recently, drawing on technical advances in production and control, the emerging kinetics of buildings and building components has been considered by Schumacher et al. (2010), who demonstrates how wide-spread such approaches have now become in the built environment. For a more generalised overview of this space, the previously mentioned framework categorises possible movement in buildings as changes of location, orientation, to building form and topology, changes to building components and to the relationship of inside and outside (Schnädelbach 2010, p. 7). Examples of movement occurring in buildings include those that offer moveable internal partitions, moveable separations of indoors and outdoors, moveable building units such as rooms on wheels, but also various types of moveable roof structures, among other possibilities.

7.2.2 Movement in People

Without exception, all expressions of human behaviour result from motor acts (Solodkin et al. 2007), and movement is therefore our main way of engaging with others and the environment. This emphasis on movements is also present in those approaches to cognition, which emphasise the embodied and embedded nature of our presence in the world (Varela et al. 1991; Wheeler 2005). Here, our bodies and the environment are seen as continuum and we cannot but leave traces in it, a fact that Richard Long has specifically explored in his environmental art (Long 1967).

Previous work that has considered human movement in the context of Adaptive Architecture focussed on the scale, expressiveness and control of movement (Schnädelbach 2016), which we briefly summarise here. The scale of movements, also used by Abawajy in their taxonomy (Abawajy 2009), ranges from micro to macro movements. This is in turn related to how visible their effect might be to an external observer: the internal movements of the cardiac muscle are relatively small scale and in many occasions invisible to others. In contrast, the muscles in our legs allow us to produce our largest movements through space and this movement clearly becomes observable. This visibility and related to this the expressiveness of specific movements are directly linked to legibility by others ('reading' someone's behaviour and their psychological state via their body movements). In the context of this chapter, human movement must be expressive to and legible by whatever sensing system is employed, as this detectability is required to make links between body movements and architectural movements. Finally, there are different levels of control that people have over their body movement. Some movements and movement patterns (e.g. breathing) are controlled unconsciously and consciously. Most of the time, people don't think about their breathing, while they clearly can for example for relaxations purposes (Montgomery 1994). Other movements are much more clearly aimed and targeted, for example when controlling fine-grained grasping actions (Solodkin et al. 2007).

7.2.3 Movement-Based Interaction

Human movement, and considerations of its scale, expressiveness and control, has been part of HCI research from the outset, while not necessarily its focus. Standard interfaces like the computer mouse and the touch screen, but also less standard technologies like head mounted displays, take human movement as input for the interaction with computing. The PUC special issue on movement-based interaction is testament to the persistent interest in this area (Larssen et al. 2007), proposing that the moving body should be considered as part of any interaction and considering the new design spaces that emerge. Over the last decade or so there have then been a number of endeavours to frame movement-based interaction design more specifically. Loke et al. have provided an overview of such frameworks in Loke and Robertson (2013) and Cruz provided an updated, more comprehensive listing recently (Ricardo Cruz et al. 2015). In what follows, we draw on these overviews to synthesize the central elements for consideration in movement-based interaction design but only as they are relevant for this chapter.

Human movement requires (a) *space*. The considered space is shaped by the human movement of interest to the designer and it might for example be sized to accommodate a single user's hand or accommodate the full body movement of multiple people. This space also becomes shaped by interactive technology, as both (Eriksson et al. 2006) and (Schnädelbach 2012) have outlined in the case of interactions supported by camera tracking and video communication, respectively. The particular (b) *interactive technology* used in movement based interaction design also influences the design more broadly. Body-worn motion sensing offers very different affordances to sensors embedded into hand-held devices or camera-based motion tracking and the relationship of sensing and interaction has for example been considered by Benford et al. (2005). At the heart of the concern is the actual (c) *interaction* to be designed in the sensed space, linking human movement to interactivity of some kind. A chosen set of body movements is sensed and used to drive a system. This might link movements of our hand to movements on screen, it might link two tangible devices together as in the inTouch prototype (Brave et al. 1998), it might amplify physiological data (Marshall et al. 2011), or it might involve whole body movements. Finally, (d) *movement in people* that can be sensed in the given interaction space and is relevant to the desired interaction is what drives that interaction. Loke et al. describe how movement has become a new design material in this context that needs to be more fully understood by changing one's own practice (Loke and Robertson 2013). In parallel, it is equally important to understand what movement can be detected best by what technology and what the key properties of available technologies are.

7.2.4 Movement-Based Interaction in Architecture

There is a wealth of movement-based interaction design in architectural history, going back nearly a hundred years. Rietveld Schröder's house (Kronenburg 2007, p. 26) offers physically adaptive features such as moveable partitions that allow the manual reconfiguration of the interior, an idea still relevant as demonstrated in Holl's Fukuoka Housing (Kronenburg 2007, p. 52). Naked House by Ban (Kronenburg 2007, p. 170) takes this a step further by providing room units equipped with wheels that can be freely placed within a larger domestic volume. Beyond entirely manual engagement and the interior, much larger scale movements have been implemented for example in Studio Gang's *Starlight Theatre (Studio Gang Architects* 2009*)* and DRMM's *Sliding House (DRMM* 2009*)*. In these cases architectural movement is driven by motors, which are triggered by consciously controlled human interactions to trigger an architectural change when desired.

In experimental architecture emerging over the last decade or so, enabled by the parallel development of applicable sensing and actuation technologies, relationships between movement in people and movement in architecture have become more subtle. TU Delft's Muscle Tower (Hubers 2004) is programmed to physically react to the proximity of people that share the same interaction space on an immediate level, Alloplastic architecture makes use of a Kinect tracker to link full body movement to the movement of a tensegrity structure (Farahi Bouzanjani et al. 2013), while Slow Furl is developed to map human presence on a much slower scale (Thomsen 2008). The more interactive couplings on this spectrum have recently been captured by work on architectural robotics (Bier 2014).

The ExoBuilding (Schnädelbach et al. 2010) and Breathe (Jacobs and Findley 2001) explorations present another take on movement-based interaction in architecture, as they make use of physiological data to actuate a physical enclosure of an interaction space. The examples across history demonstrate how human movement at various scales, expressiveness and levels of conscious control has been amplified or attenuated to be mapped into architectural movement. Despite this wealth of previous work spanning Ubiquitous Computing, Human Computer Interaction and Adaptive Architecture, there is a lack of knowledge of what such environments mean for their inhabitants. While there is continuing and growing interest to propose and develop movement-based interaction to deploy it in everyday settings, a subset of Adaptive Architecture, the growing community of collaborating architects and interactions designers lacks a full description of the design and interaction processes and the potential consequences of the designed interactions.

7.3 MOVE Platform

MOVE has been developed with the specific aim to explore the relationship of body movement and movement in Adaptive Architecture. As a platform it aims to allow for the following: (1) Physical configurability: to enable different architectural configurations (2) Mapping configurability: to enable flexible mappings between human movement and prototype movements (3) Non-expert use: by creating an interface that allows 1 and 2 to be done by non-programmers. The overall aim was to create a re-useable research tool for different contexts. In what follows, we briefly describe the MOVE platform and its development process before focusing on the interaction with one particular instantiation of the platform, prototype 3.

MOVE consists of a software platform and physical components and actuators, which can be arranged in space. Two floor-to-ceiling poles are used to mount four structural arms each, on which rotatable panels are mounted. The arms can be adjusted in height and they can be fixed in any rotated position around the cylindrical poles. Height constraints are therefore given only by a combination of panel length (considering that they might rotate toward the floor and ceiling) and floor-to-ceiling height.

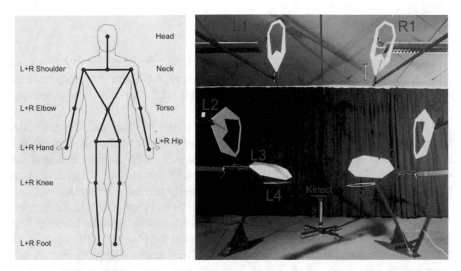

Fig. 7.1 The 15 body joints (left) as seen by OpenNI (Graphic derived from Nanoxyde GFDL (http://www.gnu.org/copyleft/fdl.html) and MOVE prototype (right) and panel layout (Top level panels L1 and R1, medium level panels L2 and R2, lowest pole mounted panels L3 and R3 and floor standing panels L4 and R4)

Panels are made from foam core board and have large cut-outs to save weight and to reduce air drag. The shape and size of the actuated components can easily be adapted. They are mounted to short rotating arms. Each panel assembly is directly fixed to the output axis of a model servomotor, which in turn is affixed to one of the mounting arms. The panels are weight-balanced to an extent to take the strain off the motor mounts. The active movement range of each panel is approximately 180°.

Using two Phidgets (Phidgets INC.) 8-channel servo controllers, four moving panels were implemented each side (with expansion to sixteen panels physically and programmatically possible). As can be seen in Fig. 7.1, panels were arranged symmetrically on the prototype, with four each side, mirroring the symmetrical nature of the human body, but not its anatomy. Asymmetrical arrangements would be possible.

A Processing Foundation (2016) project used the Phidgets21 library (Greenberg and Fitchett 2001) to interface with the Phidgets hardware and the SimpleOpenNI library to interface with the Microsoft Kinect V1. The Kinect is located at the forward centre of the MOVE prototype facing inwards to a point roughly between the two poles, but one metre out (compare Fig. 7.1). A user interface created through Swing in Netbeans allows interaction designers to make appropriate links between body movements and prototype movements, but also has the functionality to record and replay MOVE component movements. A full description of this interface is included in section 'Prototype version 3'. The functionality of the hardware and software platform allows a single person's body movement to be tracked where each of the eight panels is associated with a specific limb movement (see Table 7.1).

7.3.1   Prototype Version 1

Drawing on Borenstein et al. (2012), the first version of the implementation tracked the distance between body joints, as OpenNI sees them. As illustrated in Fig. 7.1 there are 15 points that are technically trackable and useful for interface programming, while they are not all anatomically correct.

Translating body movement to prototype movements involved calculations of the distance between two joints, the resulting dynamic value of which was individually mapped to pairs of panels. For example, the distance between the right hand and the right shoulder of a person was mapped to panels R3/L3 in version 1. On either side of the body, this version of the prototype additionally tracked the distance between elbow and torso joint, the distance between knee joints and torso joint and the distance between hand joints and the centre of mass of the respective body side. Tracking four sets of relative joint distances allowed the mapping to the four sets of two panels, so that the entire prototype could be actuated. Table 7.1 describes the particular, chosen mappings in the context of all Kinect trackable movements and available MOVE movements. A person's shoulder pitch is mapped to panels L1 and R1, shoulder yaw is mapped to panels L2 and R2, Elbow pitch is mapped to L3 and R3, while Knee pitch is mapped to L4 and R4.

Table 7.1 Mapping of the degrees of freedom (DoF) of the upper and lower limbs of a person (feet and hands are assumed rigid here) to the degrees of freedom available in MOVE (drawing on Herman 2007). The 26 degrees of freedom available on the left and right hand side of the body are mapped to the eight degrees of freedom available in the MOVE prototype L1-4 and R1-4

Upper Limbs (Arms) 7 DoF total	Shoulder Ball + Socket 3DoF			Elbow Hinge 1 DoF	Forearm Pivot: Una Radius 1DoF	Wrist Ellipsoidal 2 DoF	
Tracked Body DoF	**Pitch**	Roll	**Yaw**	**Pitch**	Roll	Pitch	Yaw
Mapped MOVE DoF	**L1 + R1**		**L2 + R2**	**L3 + R3**			
Lower Limbs (Legs) 6 DoF total	Hip Ball + Socket 3DOF			Knee Hinge 1 DOF		Ankle Saddle Joint 2 DOF	
Tracked Body DoF	Pitch	Roll	Yaw	**Pitch**		Pitch	Yaw
Mapped MOVE DoF				**L4 + R4**			

We evaluated this first functional version of MOVE in a demonstration and focus group session (lasting around 90 min) with five colleagues from our lab, to feed information forward into the development process. A short presentation was followed by a demonstration and a session in which each of the participants tried the prototype. Unstructured discussion during this trial was followed by a recorded semi-structured focus group. A first enthusiastic response from participants and statements about empowerment (the prototype made one participant feel bigger) was tempered by concerns about ethical data use and privacy. In particular, the 'nervous' nature of this first implementation (the servo motors always being ready to engage and slightly shaking) would have made people uncomfortable during longer spells of use. The mapping between body and prototype was also not seen as accurate and predictable enough. Participants requested a better representation of the body movements mapped to the prototype as they found them hard to comprehend. The discussion included suggestions for additional mappings, for example the reversal of directions of movements from body to prototype, and a better interface to the possible mappings, including all configuration options. With regards to applications, participants discussed how the prototype would be useful to training in martial arts and dance and may be theatre where the audience could have input too.

7.3.2 Prototype Version 2

Further development focussed on addressing the tracking issue. Rotational sensing was implemented where the angles between limbs were tracked, instead of the distance between joints. This provided for improved tracking accuracy and stability in practice. Four participants (P1-P4) were recruited at University to experience an introduction to MOVE, a short demonstration and an individual trial session with the prototype. This was followed by a brief semi-structured interview. There were males and female, all between 20 and 25 years old. We briefly summarise the resulting formative feedback.

Movement—Understanding: We observed how participants moved to acquaint themselves with the prototype, movement itself being the learning strategy. They tried things out, seeing MOVE react, adjusting to what it can do, and for some participants this translated into a form of exercise. Participants requested more trackable movement types, for example the idea to track and adapt according to the torso direction was mentioned.

Connectedness, Speed—Delay: Perceived delays and comparatively slow speeds of the prototype meant that only certain movements gave participants a sense of connectedness. We observed how this could lead to adaptations in behaviour, so that participants moved in ways that they had learnt could be tracked by the prototype.

Inaccuracies—Jitter: A certain jitteriness of the prototype is clearly noted, caused by the rapid cycle of reaction and counter reaction in the digital servos used, which keeps those in position and ready for engagement. Inaccuracies in tracking are also evident to participants, as the sensor does not handle self-occlusion or rotations away from frontal view very well.

7.3.3 Prototype Version 3

For the final prototype version that was then studied, we replaced all remaining digital servomotors with analogue alternatives. This simple change in hardware reduced the jitter in the prototype considerably, and the slight loss in reactiveness of the engines was not substantial.

Beyond this smaller change, refinements included experimentation with more trackable movement types, such as the distance between the two arms, the distance between the two legs, hip pitch and torso rotation. While these were not used in the trial of the MOVE prototype described below, they prompted the creation of a more complete design interface to the movement mapping available in the prototype, as shown in Fig. 7.2. The interface allows the eight trackable body movements to be mapped to the 8 (possible) pairs of servomotors, giving access to control over their velocity, acceleration as well as their range.

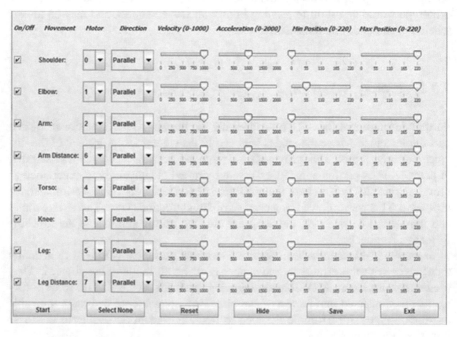

Fig. 7.2 MOVE interface to mapping possibilities. The interface lets designers map body movements to prototype movements, allowing adjustments to panel velocity, acceleration, as well as min/max positions. Configuration can also be saved here

7.4 MOVE in a Martial Arts Context

For a trial of the final version of the prototype, we invited two Tetsudo practitioners in two sessions. Tetsudo is a martial arts, which emerged in the 1960's, drawing on other martial arts forms (Dhaliwal 2016). It focuses on the unity of body and mind and emphasises self-control over self-constraint. This results in the aim to avoid being angry and aggressive when there is no need and avoiding to be passive and hesitant, when there is a need to act.

In practice, Tetsudo is contact-less and has a substantial performance aspect to it, enacted through Tetsudo 'Kheds'. These are '...*set pieces of 'imaginative conflict theatre, in which the artist immerses himself/herself to experience the full array of the physical, emotional and intellectual dimensions of a 'conflict situation.'* Dhaliwal (2016). Teaching is conceptual, rather than prescriptive, so that no two Tetsudo practitioners perform Kheds in exactly the same way. Both individual performance and sparring with a partner are important parts of the practice. We invited Tetsudo practitioners because of their focus on own body movement, movement skills and expected ability to reflect on body movement when providing us with feedback.

7.4.1 Method

The Tetsudo practitioners (P5 + P6) participated in two trial sessions. P5 attended the first session (Tetsudo 1) alone. He has practiced Tetsudo for around 8 years and wears a purple belt. P5 joined P6 in the second session (Tetsudo 2). Her experience is of more than 15 years and she wears a double black belt. The two experimenters were present throughout but mostly out of sight of P5 and P6. Sessions were recorded with 3 video cameras, one positioned at the back, one at the side and one at the front. A semi-structured interview was recorded with P5 following his session and with P5 + P6 following their joint session. Video and audio were transcribed across the sessions and interviews. We draw on these transcriptions for the descriptions of the session structure and for the development of analysis themes, which are used below to describe the interaction with MOVE in the context of Tetsudo.

7.4.2 Overall Session Structure

The experimenter introduces the mapping of body movements to each pair of panels, moving from L1R1 to L4R4. Following this, the participants are asked to complete four MOVE postures: Close all panels, open all panels, close right hand side and close left hand side (see Fig. 7.3). Both participants complete this warm up period in approximately five minutes without major difficulties or concerns and it provides

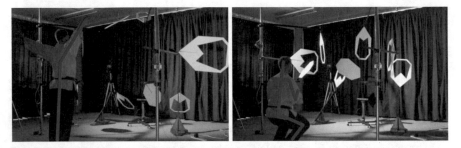

Fig. 7.3 MOVE—Open and closed with the corresponding body postures by the participant

them with a base understanding of the prototype capabilities and how to manipulate MOVE with their bodies.

These short introductions are followed by individual 'free' sessions, in which both participants explored Tetsudo movements and postures in relation to the prototype. Participants 5 and 6 spent around 20 min and 10 min respectively in those individual free sessions. The character of the two individual sessions was different, in that for P5 it was their main activity during their trial session. For P6, it was the first half of the session, leading into a shared part (P5 and P6 working together; see section Shared session—P5 and P6 below). However, P5 and P6 both had sufficient time with the prototype by themselves to get fully accustomed to it. Deliberately, no detailed instructions of what to do or achieve were provided for the free session by the experimenters. As for P1–P4, the aim was to let them explore the possible interactions with MOVE, but this time specifically framed by their Tetsudo experience and their learnt movement repertoire.

7.4.3   Individual Session 1–P5

The individual free session of P5 begins with a bow, the same way that a Tetsudo session starts (Fig. 7.4). P5 continues the session with a wide-opening 'ready' posture, as illustrated below. Standing still, both arms move up and side-ways to come back down again in front of P5. This movement is particularly well tracked. P5 returns to a short pause and this set of movements in this session, whenever tracking has been poor, or when to begin trying something new (Fig. 7.5).

The first part of the session is characterised by a quite methodological exploration of what postures P5 can achieve working with MOVE, referring both to postures that MOVE can exhibit and those that P5 can demonstrate. This includes keeping still, with the expectation that the prototype would remain still. When it does not, this is being noticed and being checked, as illustrated below, when L4 continues to move although P5's left leg remains stationary.

P5 also trials upper body movements standing on one leg, crouching down, moving arms synchronously and asynchronously, crossing limbs over and keeping them separate and facing forward as well as away from the Kinect sensor, Tetsudo kicks and punches, executed passively (slow) and vigorously (fast). For the last three minutes of the session, P5 concludes with a section using Tetsudo sticks. At this point, interaction becomes most deliberate, possibly framed by the constraints the sticks put on the interaction. P5 crosses arms over less frequently.

Fig. 7.4 Bow following warm-up and at start of free session

Fig. 7.5 Sequence of Tetsudo ready posture with P5 raising arms and MOVE open (left) and P5 with completed ready postures and MOVE mostly closed (right)

7.4.4 Interview

During the interview P5 began to reflect on the movement relationship to MOVE, stating that there:

> ... were some moves, which would kind of bring it all together ... . There was certain postures ... it was very reflective of that ... . We do something called a ready posture ... you start almost everything with this. It was nice to feel that responsiveness. (compare with Fig. 7.5).

P5 continues to describe the lack of responsiveness in other circumstances when he reflects on the speed with which MOVE can track a performer:

> It is a little linear ... and I think the pace, the speed is ..., if I did things rapidly, it couldn't respond to it ... . So I stuck to moves that it was responding to, ...

P5 reflect on the fact that this might suit a Tetsudo beginner quite well, as students start out repeating quite linear movements. However:

> ... when you get more ... advanced, it's much more kind of around ... the circular movements and spinning kicks and things like that and I think it [MOVE] would ... possibly not respond so well to that.

7.4.5 Individual Session 2–P6

The second individual Tetsuo session had quite a different character from the first. P5 invited P6 to join in the second trial, roughly five weeks after the first trial, proposing P6 because of her superior Tetsudo skills. In the intervening period, P5 would have had the opportunity to explain the prototype from his perspective and his interaction with MOVE. Such explanation is clearly present in the individual session of P6, when P5 first proposes to do Khed movements passively *and* vigorously during the session, but then adds: '... it [MOVE] struggles with ... the speed ... it manages to cope better, if you are moving slowly.' While P5 stresses that there is no right and wrong and proposes P6 explores the whole range of Tetsudo, P5 clearly frames the MOVE range of capabilities for P6. P5 remains in the space with P6 during this second individual session, outside tracking range, and P5 and P6 occasionally discuss during the session and also involve the experimenters when they have specific queries (Fig. 7.6).

Fig. 7.6 P6 showing fine-grained control over MOVE in a number of set poses following each

P6 begins the session with a broad range of Tetsudo moves, before a short focus on leg kicks. Both passive and vigorous Khed movements are visible, with the faster movements showing considerable delay in the MOVE response. About seven minutes into the 10-minute session, a sequence of Khed movements lasting for around 50 seconds, best demonstrates the level of control that can be achieved working with MOVE. The fact that control has been achieved is confirmed by P6: *'Feel like, I have got them (the panels) tamed now.'*.

7.4.6 Shared Session–P5 and P6

Following the individual trial by P6 during the Tetsudo 2 session, both P5 and P6 used MOVE in a joint session, which lasted for 26 min in total. Both participants were in control for parts of the session, with the respective other participant standing on the opposite side behind the Kinect sensor, joining in or just observing (see Figs. 7.7 and 7.8). The session was interspersed by discussion, P5 and P6 discussing what they wanted to try out, occasionally confirming with the experimenter. Broadly, the session can be split into three parts: During a first part, Tetsudo freestyle sparring lasted for approximately ~7:30 min, with P6 being tracked. During the middle part lasting ~6:50 min, the session is characterised by trialling expressive postures. In a final part, lasting for ~9 min, P5 and P6 mirror each other with MOVE mirroring the lead participant. The three parts are further described below.

Fig. 7.7 P6 Lurching forward towards P5 (left) and Move following, pointing at P5 (right)

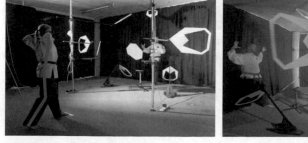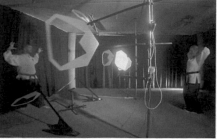

Fig. 7.8 P6 creating intimidating MOVE posture (left) and P6 'hissing' at P5 through MOVE (right)

7.4.6.1 Freestyle

Tetsudo distinguishes between compromised, competitive and combat freestyle, successively increasing the level of contact with the opponent and the force of the contact (Dhaliwal 2012). All are with partners. Compromised is aimed at helping in training, making use of Kheds learnt and sequencing those. Competitive (light contact) and combat (firm contact) are scored in competitions. P5 describes freestyle as 'action and reaction', with one person reacting to opportunities to strike, which the opponent created by leaving open space for attack. Both action and reaction movements are aligned with a set repertoire of recognised movement sequences or Kheds. In the first part of the joint session, MOVE mirrors P6 with P5 engaging from around 3 m away, standing behind the Kinect sensor, as described earlier. Using MOVE, freestyle here seems to be most similar to compromised as a large gap is introduced, i.e. there is no contact between opponents.

During the first two minutes, freestyle sparring proceeds at regular 'Tetsudo' pace', which is generally too fast to be tracked accurately by MOVE. Prototype movements appear quite erratic and sparring seems to ignore the prototype in some sense, with participants fully concentrating on the sparring partner. This prompts P6 to suggest: '*Shall we go slower, do you think ...*' with P5 agreeing to that suggestion. During the two minutes that follow, freestyle sparring continues at a considerably slower pace, with much better movement mapping by the prototype, and P6 confirms this by stating: '*I am working on making my reaction to P5 much more clearer, rather than too complex.*'.

From an observer's viewpoint, sparring now appears much slower and the interaction with MOVE is becoming part of the Tetsudo routine in a way that seems to 'make sense' to both participants. However, MOVE adds something to the Tetsudo practice, which P6 describes during a brief in-session discussion as follows:

> I am fighting P5, well or working with P5, with a distraction (pointing at MOVE) ... not so much a distraction actually, with an extra little thing to think about. It's like ... you are doing a dance and saying your five times table or something ... You know, doing two things, a physical thing and then an awareness thing. And ... I quite liked it.

The above highlights two distinct influences of the MOVE prototype on the performers' practices. First, they adapt the speed and 'clarity' of Kheds to allow MOVE to follow them, which has a slowing effect and it makes movements more legible. In addition, for an experienced performer, MOVE seems to add an extra layer of complexity to their movement practice. This complexity results in a worthwhile challenge for P6 and a possible extension to the Tetsudo practice per se, challenging a person through two concurrent embodied interaction patterns.

7.4.6.2 Expressive Postures

A brief discussion at the end of the sparring session results in P6 experimenting with making MOVE more clearly part of their 'attack', drawing the panels into be part of

Fig. 7.9 P5 mirroring P6—Stretched out (left) and crossed over (right)

the action by P6, which P5 would have to respond to (during this short episode, P5 is only observing). Following this, P6 lurches forward in a throwing movement (see Fig. 7.7 left), which result in the MOVE panels lurching forwards towards P6 (see Fig. 7.7 right), albeit at reduced speed.

This results in P5 stating (pointing at panels L2 and R2): '*Yeah, these two are intimidating.*' and P6 continuing with another noticeable body posture as illustrated in Figs. 7.8 and 7.9. This is commented on by the two participants:

P6:   *That's everything up.*
P5:   *I mean, it's, in terms of uhm, that, what you had it just then, it's, it's quite a*
        you know, powerful sort of gesture ... that kind of like the way you had it then.
P6:   *Yeah*

This posture is carefully repeated by P6 before the conversation continues. P6 structures her body posture for maximum effect in the MOVE posture, re-enforcing the creation of an intimidating expression, while her own posture is not necessarily intimidating. P5 reacts to MOVE's expression with an expressive posture of his own, raising both arms and spreading out all fingers. He states: '*Yeah kind of like ... (making an aggressive hissing sound)*'.

This demonstrates the potential for people to amplify their body postures through MOVE in an expressive way and the ways that this is perceived by a counterpart as expressive. It also shows a deliberate strategy of making the prototype 'fight' on one side, with one of the participants opposing the other.

7.4.6.3   Mirroring

At the beginning of part 3 of the shared session, P5 takes over to be tracked. He demonstrates his mastery over the prototype through controlled movements, resulting in very good tracking. Initially observing only, P6 then begins mirroring P5, before suggesting that they continue the session in this fashion. This is agreed and it leads to a three-way connected system: P5 exhibiting relatively slow, deliberate Tetsudo movements, with MOVE and P6 mirroring those movements (see Fig. 7.9). From the observation it seems clear that both parties know the movement sequences, which makes it possible for P6 to mirror P5 well. A roughly two-minute section of this is followed by a brief discussion of symmetry, with P5 expressing that he thought that symmetrical body movements are more successful than moving only one side of the body. Participants then swap over for P6 to be tracked, with mirroring continuing, exhibiting both one-sided and mirrored body movements, confirming this in a brief discussion:

P5:   So, where you doing it the opposite, so if I was doing it with my left hand (raising left hand), you were doing it with your right?
P6:   Yeah.
P5:   I just mirrored everything that you did

This second mirroring sequence produces some of the calmest and best-tracked sequences with clear evidence that P6 fully understands the tracking range of MOVE.

This section concludes Part 3 of the joint Tetsudo 2 session. It demonstrates how MOVE can be used as a dual-mirror, it following one participant and the second participant following the first and MOVE itself.

7.4.7   Interview

The semi-structured interview following the shared session probed participants' reactions to the prototype, to the appearance of the prototype, desired extensions, possible applications and possible opportunities and concerns. In what follows, we briefly reflect on the three core themes emerging from the interview: interaction, limitations and uses in the Tetsudo martial arts practice.

7.4.7.1   Interaction

The participants discussed how a moving environment like MOVE would draw people into explore more, out of curiosity. This is because encountering MOVE presents the challenge of working out what connections there are and how they work; what moves in relation to what and establishing a link to it as P6 explains:

I suppose it's trying to make that link between something that's totally separate to you.

Reflecting comments made during the trial session, P6 also re-iterates the fact that MOVE presents a challenge once a link has been established:

… it's really fascinating to have this extra dimension that you're controlling. It challenges a different part of your brain when you're moving.

The fact that there is reaction of some kind is what supports the establishment of a link between participant and MOVE, while the particular appearance of the prototype is not important in this as expressed by P5:

I think the physical appearance of this is not important. More so the fact that it's responding to the movement.

Beyond visual feedback, sound is mentioned as a contributing factor to creating a specific connection with the prototype, as explained by P6:

I think when you've got an audible link with the movement it instantly adds another connection with it. So, if you took that sound out and it was completely silent … there'll be another level of connection, whether it will be a deeper level of connection or a lighter level … It's the senses, isn't it? It's how you're sensing that piece of equipment.

Through this comment, participant 6 highlights the importance of taking a holistic view in the design of feedback environments.

7.4.7.2   Limitations

The two participants agreed that they were looking for more familiarity of the prototype using it first, and that this might have delayed making a link with the prototype initially, and P6 states:

I think we're always trying to make something familiar, so you're trying to make that look like two shoulders, two elbows. And that's why I feel that we struggled with the knee one [referring to panels L4 and R4].

Neither of the two participants directly suggest a more anthropomorphic appearance of the prototype, but something where mappings are more easily legible and where those mappings more directly reflect a person. The above also points to the fact that the movement range of the prototype was seen to be too limited. While L4 and R4 where mapped only to the movement in the knee, participants expected those panels to more closely follow the movement range of the whole leg, which would have enabled the mapping to kicks, for example. The two participants continue to discuss this more specifically in relation to Tetsudo. In particular, the lack in variation in speed was mentioned. As highlighted earlier, Tetsudo includes rapid and slow movements and MOVE cannot express both as P5 outlines:

… from a Tetsudo perspective, … we have the sort of … range of … styles from very passive movements to very vigorous movements, and at the moment I think it only really responds to the very passive, slow movements.

This results in vigorous Tetsudo movements (combining strength and speed) not to be represented at all, providing limited expressiveness. A simpler, but very important limitation was the participants' expectation for the prototype to be still, when they perceived themselves to stand still.

7.4.7.3   Role in Martial Arts

The final theme concerned the risks and opportunities of the application of such a prototype in martial arts. With the focus on the visual and external, participants saw a risk in loosing the focus on one's own body, which is essential to Tetsudo training, as P6 stated:

> ... where you're putting the onus on something more visual to give you, to help you with your movements, your physical body movements, I think you're taking away something. You're taking away mobility that this one human being normally does ...

This is discussed as part of the more general observation that people are increasingly dependent on technology in all aspects of life. At the same time, a prototype such as MOVE would have great potential during the training of Tetsudo as P5 concludes:

> ... as a part of training, I think something like this would be a really interesting thing for someone to do ... when they're learning Tetsudo ... It would become an integral part of the training.

7.5 Summary of Findings and Discussion

Kinetic Adaptive Architecture is increasingly being built around us, as evidenced in the introduction and background section. Within this field, movement in architecture can be related to movement in people and a smaller set of prototypes are now proposed that have this interactional capability, providing new forms of interaction. This chapter has drawn on its broad, multidisciplinary context and the iterative development and evaluation of the MOVE prototype. This lets us now reflect on how the feedback loop between person and environment shapes and limits interaction, and how the selectiveness of the MOVE 'mirror' becomes useful in practice and training. We conclude with an outline of the co-creation process in this context: (1) Designers of movement-based interaction embedded in Adaptive Architecture need to draw on and design around the correspondences between person and environment. (2) Inhabiting the created feedback loops result in an on-going form creation process that is egocentric as well as performative and embodied as well as without contact.

7.5.1 Feedback Loop Between MOVE and Body Movements

The movement range of MOVE is clearly limited in comparison to human movement and this is also evidenced throughout the trial feedback. The Kinect sensor does not track some presented movements. For example, the rotational movement in the lower arm is not being seen by the Kinect, and hands and feet are seen as solid. As demonstrated in the trial sessions, Tetsudo includes untrackable movements such as finger movements, but it also includes untrackable relationships of body movements, such as the crossing of limbs. Tracking technology continues to evolve, and the use of Kinect version 2 would have resolved some of the above issues. However, there will be tracking technology limitations for some time to come, which means that human movement tracking will remain incomplete in practice.

There are only 8 (4 pairs) of single degree of freedom movements that MOVE can produce, even though they appear to be much more complex when executed in unison. Given that feet and hands are seen as solid by the Kinect sensor used, participants are presenting 26 trackable degrees of freedom to the camera (13 on each side (compare Table 7.1)), when facing forwards. Unlike human movements, which are fundamentally integrated with each other (Bernstein 1967, p. 22), there is also no interrelation between the movements that MOVE can produce, all panels being actuated and mapped independently from each other. Whereas the two Tetsudo performers deliberately varied the speed of their interaction, as it is part of their practice, the MOVE prototype only coped with the slower and less forceful movements. When fast movements were followed by a short pause, MOVE was able to catch up. Through a sequence of fast movements, MOVE was not able to keep up. The experienced slowness results from the overall system performance chain, including tracking performance, motor acceleration and speed, as well as panel inertia.

The differences in appearance, the tracking issues, lower number of degrees of freedom and slower speed of MOVE in comparison to a person, result in participants and prototype having very different embodiments and capabilities. An anthropomorphic robot was never under consideration for this work, as the emphasis was on the investigation of the relationship of human and architectural movement. For this reason, overcoming the correspondence problem (Nehaniv and Dautenhahn 2002) was not a specific aim (and it seems clear that solving the correspondence problem is currently not possible, even if that was desired).

Importantly, the lack of correspondence in embodiment between participants and MOVE does not lead to a breakdown of the experience or the abandonment of Tetsudo in this particular context. Instead, the practice is subtly adapted, even during the relatively limited trial periods described here. Following an exploration by participants of MOVE's tracking, mapping, range and speed and the associated learning, the two Tetsudo experts perform a version of their practice which concentrated on slow, passive movements, avoiding the cross-over of limbs, while facing forwards into the camera. A practitioner with the right level of experience in controlling their body movement adapts to the range of postures and speeds that MOVE can perform. This observation confirms Nehaniv's assertion that even without a match in embodiment between 'demonstrator' and 'imitator', correspondence between the two can still be recognisable (Nehaniv and Dautenhahn 2002).

In the above sense, MOVE acts as limiter in a person's movement range. While a performer could continue performing as if MOVE was not there, the experience is visibly much less rewarding, when they do. Then, MOVE moves when it should not or simply ends up in the 'wrong' place at the wrong time. When the performer understands the limitations, the experience does become rewarding. In this way, the architectural prototype and the performer have been mutually incorporated, to draw on a concept from social interaction (Fuchs and De Jaegher 2009). This interaction would allow an experienced practitioner a focus on the individual movements that they need to retrain or it might be fine-tuned to challenge a performer in specific ways.

7.5.2 A Selective Mirror

Given the lack of embodiment correspondence described above, the study demonstrated how MOVE acts as a three-dimensional, selective mirror for the person being tracked. It reflects the movements of a person back to them in a particularly structured way. This is relevant especially as mirroring and imitation are embedded in the movement arts (e.g. dance, martial arts). In dance, the interior architecture is often designed to support the practice by providing large mirrored surfaces in the dance studio. The mirror allows students to reflect on their own performance, while the performance of other students and the teacher also become visible. Beyond this, Architecture includes mirrors in other people-centred circumstances such as children's play (ArchDaily 2012) and performance art (Kohlstedt 2004).

Digital mirroring also finds its use in dance and motion based analysis is now common in sports more generally, as Barris et al. review (Barris and Button 2008). The YouMove training system uses a half-silvered mirror and Kinect-tracking to measure to what extent people are following prescribed movement patterns and to help with retention (Anderson et al. 2013). In early work around dance, Hachimura et al. use a motion tracker to re-create a dancer's movements through a 3D avatar (Hachimura et al. 2004). More recently, Kyan et al. deploy Kinect tracking within a VR projection to allow dancers to review their own performance with the help of a mapped avatar (Kyan et al. 2015).

Beyond physical mirrors in the environment and digital mirroring, there is an accompanying form of mirroring, both in dance and martial arts, in that students mirror and eventually imitate their teacher. Tetsudo can serve as a useful example here. It has a theoretical repertoire of movements that is taught by demonstration, mirroring and imitation. Each performer has an individual range within that theoretical repertoire and they express that range through their bodies in particular, performed instances. The movement ranges are not the same for two performers, even when comparing performers at the same 'level'. In a typical Tetsudo class, the teacher will face a group of students, performing a series of movement routines, which the students will observe, mirror, learn and then be able to imitate without the teacher in front of them. Over time, the routines can become part of the students' own movement repertoire. This involves the concentration and determination of the students and reflection on their own movements as the interviews and the video evidence have shown (see Fig. 7.6, P5 checking down).

In parts of the shared session between the two Tetsudo performers, it was then possible to observe the combination of the two forms of mirroring, outlined above: the environment mirroring one performer, and one performer mirroring another (via direct observation and via the environmental mirror) (compare section 'Shared Session - Mirroring'). This is similar to the traditional dance theatre set-up apart from the fact that MOVE is of course not a mirror at all, as it breaks down the performed movements and reduces them to eight degrees of freedom that it can express. Some movements are simply not mirrored and what is mirrored therefore becomes amplified in the feedback loop. This selective mirroring and giving practitioners the choice over what is mirrored is a key property of MOVE that can be exploited in teaching, through the attenuation or amplification of what movement becomes mirrored, guided by the teacher or indeed the student. While this would be possible in screen-based (projected or head-mounted) alternatives, mirroring MOVE is physically highly immersive and performative. The Tetsudo performers agreed that a movement prototype like MOVE would be a useful tool for a martial arts beginner, as training is characterised by a focus on simpler, as of yet unconnected movement sequences. Breaking down and slowing down the movements to be learnt will actually help students in their learning.

7.5.3   *Co-creation: Associating Human and Architectural Movements*

Drawing on the broader context and previous work in this space, it is then possible to describe the embedding of movement-based interaction into the environment more specifically and how this leads to the co-creation of kinetic adaptive architecture. In Schnädelbach (2010) the general relationship between human behaviours and architectural behaviours has been sketched out. While making associations between human and architectural movements is principally included in this work, it was not its focus and was not described any further. Work with ExoBuilding (Schnädelbach et al. 2010) has lead to the description of the feedback loops that emerge between people's behaviour and behaviours in adaptive architecture (including movement behaviours) and those remain at the core of architecture that is adaptive to people. Specifically focusing on movement, human movements as they are relevant for the mapping to architecture have then been described, including an overview of movement control and detection (Schnädelbach 2016). In this work, a number of examples from architectural history was used to begin to set out movement associations along the three axes of scale, expressiveness and control as well as the polyrhythmic relationships between human behaviours and environment as highlighted by Lefebvre (2013).

The MOVE platform was a direct response to the theme developing. It was specifically created to investigate making links between human and architectural movement. The study of the first prototype built with the platform has then delivered key insights to describe the co-creation process of movement-based interaction in Adaptive Architecture.

7.5.3.1 Creating Feedback Loop Through Correspondences—Opportunities and Constraints

When associating human movement with movements in the environment, the core concern of the architect or experience designer is to create the feedback loop between built environment and people. Creating this feedback loop involves making correspondences between the movement behaviours of people and the movement behaviours of the environment and this is evident in work across Adaptive Architecture such as Hubers (2004), Farahi Bouzanjani et al. (2013) and Bolbroe (2013). As we have seen in the study feedback, being selective about what becomes mapped to what is essential in this. What might correspondence here mean? The least correspondence would be a static environment, within the larger confines of environments never being completely static (refer to section Movement in Architecture). This would not present Adaptive Architecture in the sense that it is discussed here. The greatest correspondence would be a clone of a person (exactly the same embodiment) that somehow follows all movements of that person. This is clearly not possible but also does not concern Adaptive Architecture.

Within this extensive range of theoretical correspondence relationships, creators consider correspondences between environment and people with regards to spatial relationships, temporal relationships and control. Spatial relationships are concerned with mapping form, scale and degrees of freedom in movement between environment and people. Temporal relationships are concerned with mapping the tempo of movements between environment and people to create isorhythmic, polyrhythmic and arrhythmic relationships to use Lefebvre's thinking (Lefebvre 2013). Finally, the architect needs to consider autonomy of the environment, i.e. can it move by itself or is it only reactive to the person inhabiting it, for example being able to lead a person rather than just follow it.

Creating anthropomorphic robots and therefore very close correspondences between people and environments was not the aim of this work, and when architecture is concerned it probably rarely will be. Also, as has been pointed out earlier

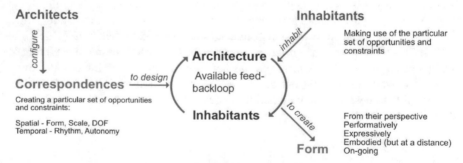

Fig. 7.10 Developing Movement-based interactions within Adaptive Architecture. Architects configure correspondences between people and the environment to create feedback loops. Inhabitants inhabit those feedback loops to create egocentric form, performatively and expressively, in an embodied but distant and on-going fashion

above, people readily perceive correspondences, even if form and scale for example are very disparate between to items observed (Nehaniv and Dautenhahn 2002). In other words, architects and experience designers can achieve recognisable feedback loops between environments and people based only on a few points of correspondence. They have achieved this by creating a specific set of interaction opportunities and constraints (Fig. 7.10).

7.5.3.2 Creating Form and Feedback Loop

A person inhabiting the resultant Adaptive Architecture is faced by those opportunities and constraints to link their movements to those of the environment. In that sense, the inhabitant's control over their environment is very much specified by someone else, for example describing exactly how particular body movements relate to particular environmental movements. The experimental work described here, has then demonstrated how people, through engaging with the feedback loop created by the designer, create form. The prototype gives people an unusual and on-going level of agency in shaping their environment: through body postures, architectural elements are arranged in space in a fluid and immediate way. This form creation process has a number of key properties:

Form creation occurs from an inhabitant perspective. It is created very much from the inside, from a central viewpoint in the environment to be created. In this, it appears most similar to design processes in Virtual Reality (Bourdot et al. 2010) allowing experts and non-experts to consider architectural designs in an immersive fashion. It might also be related to the perspective a crafts-person takes when measuring and laying out a room to be re-fitted, standing in the room during the design process. It is most dissimilar to the processes employed by architects, who view their designs mostly from the 'outside', and who produce detailed representations (e.g. drawings and models) before implementing a design. Related to the above, but not just considering the design process, the resulting architecture is fundamentally 'egocentric'. Any form created by the tracked person, for example by aligning sets of panels up in particular ways, only 'makes sense' from their view point. Architecture is frequently concerned with framing views for inhabitants but the observer has to move into the correct position to appreciate this design. With MOVE, the inhabitant remains relatively stationary and arranges the environment around them to create form and to frame their view.

At the same time, this creation of form is very performative and presents an opportunity for expression. The interaction of the architect-inhabitant with their environment is clearly visible by others in the same space. In most observed cases during our studies, this form is not created for an audience and their point of view, and therefore the created form will be illegible, or at least it will appear very differently from a public viewpoint. A special case was observed when a second performer was deliberately added to the interaction, and MOVE was employed to re-enforce a threatening expression. This demonstrated how the expressiveness in the prototype could be used to amplify expressions or indeed to attenuate them. Finally, there is a peculiar absence of actual contact with the environment for something that looks and feels very much embodied to the tracked, interacting person. The coupling of body movement and panel movements is at a distance. This is very different from similarly constrained Adaptive Architecture (with regards to the range of overall movements it can produce), such as the aforementioned Schröder house (Kronenburg 2007, p. 26). It is also different from the slower process of creating form that a craftsperson employs, handling physical elements, assembling them in space, adjusting and fixing components where they belong. Most importantly, the form creation process is on-going, with the architectural form continuing to adapt, as long as the person is engaging with it. People are at the absolute core in the type of Adaptive Architecture described here.

7.6   Conclusion

The research presented in this chapter has synthesized previous work with the development and evaluation of the MOVE prototype. MOVE is a dedicated research platform to explore movement-based interaction in Adaptive Architecture. We have discussed the ways in which the feedback loop emerging in MOVE shapes and limits interaction and how selectively mirroring interaction can be employed. Drawing on the above, we have described the co-creation process that is enabled: (1) Architects and designers of motion-based interaction in Adaptive Architecture create the feedback loop for a specific application area, focusing on correspondences between people and the adaptive environment, (2) Inhabitants map their movements to architectural movements within the existing set of constraints to create form on an on-going basis.

Acknowledgements We would like to thank all study participants and the two Tetsudo performers in particular. This research has been funded by the University of Nottingham through the Nottingham Research Fellowship 'The Built Environment as the Interface to Personal Data' and the EPSRC via grant EP/M000877/1 'Living with Digital Ubiquity'.

References

Abawajy JH (2009) Human-computer interaction in ubiquitous computing environments. Int J Pervasive Comput Commun 5:61–77

Anderson F, Grossman T, Matejka J, Fitzmaurice G (2013) YouMove: enhancing movement training with an augmented reality mirror. In: Proceedings of the 26th annual ACM symposium on user interface software and technology. ACM, St. Andrews, Scotland, United Kingdom

Archdaily (2012) Mirror house/MRLP. ArchDaily. http://www.archdaily.com/200735/mirror-house-mlrp. Accessed 10 Mar 2016

Banham R (1984) The architecture of the well-tempered environment. University of Chicago Press, Chicago

Barris S, Button C (2008) A review of vision-based motion analysis in sport. Sports Med 38:1025–1043

Benford S, Schnädelbach H, Koleva B, Anastasi R, Greenhalgh C, Rodden T, Green J, Ghali A, Pridmore T, Gaver WW, Boucher A, Walker B, Pennington S, Schmidt A, Gellersen H, Steed A (2005) Expected, sensed, and desired: A framework for designing sensing-based interaction. TOCHI 12:3–30

Bernstein NA (1967) The co-ordination and regulation of movements. Pergamon Press, Oxford, New York

Bier H (2014) Robotic Building(s). Next Gener Build 1:83–92

Bolbroe C (2013) Adaptive architecture. Copenhagen, Denmark. http://adaptive.itu.dk/projects/A daptive_Architecture/. Accessed 26 Sept 2017

Borenstein G, Odewahn A, Jepson B (2012) Making things see: 3D vision with Kinect, Processing, Arduino, and MakerBot. O'Reilly, Make Books, Sebastopol, CA

Bourdot P, Convard T, Picon F, Ammi M, Touraine D, Vézien JM (2010) VR–CAD integration: multimodal immersive interaction and advanced haptic paradigms for implicit edition of CAD models. Comput-Aided Design 42:445–461

Brand S (1994) How buildings learn: what happens after they're built. Viking, London, UK, New York, USA

Brave S, Ishii H, Dahley A (1998) Tangible interfaces for remote collaboration and communication. In: Proceedings of the 1998 ACM conference on Computer supported cooperative work. ACM, Seattle, Washington, USA

Bullivant L (ed) (2005) 4dspace: interactive architecture. Wiley-Academy

Coyne R (2016) Places to play. Interact Forthcom

Dhaliwal GUBS (2012) Introduction to Tetsudo. In: Association T (ed). Tetsudo Association, UK

Dhaliwal GUBS (2016) Tetsudo-the art form. http://www.tetsudo.co.uk. Accessed 10 Mar 2016

DRMM (2009) Sliding house. http://drmm.co.uk/projects/view.php?p=sliding-house. Accessed 03 Mar 2015

Duffy F (1990) Measuring building performance. Facilities 8:4

Eriksson E, Hansen TR, Lykke-Olesen A (2006) Movement-based interaction in camera spaces: a conceptual framework. Pers Ubiquitous Comput 11:621–632

Farahi Bouzanjani B, Leach N, Huang A, Fox M (2013) Alloplastic architecture: the design of an interactive Tensegrity structure. In: *Proceedings of the 33rd annual conference of the association for computer aided design in architecture (ACADIA)*, Cambridge

Fox MA, Kemp M (2009) Interactive architecture. Princeton Architectural Press, New York

Fuchs T, de Jaegher H (2009) Enactive intersubjectivity: participatory sense-making and mutual incorporation. Phenomenol Cogn Sci 8:465–486

Greenberg S, Fitchett C (2001) Phidgets: easy development of physical interfaces through physical widgets. In: 14th annual ACM symposium on user interface software and technology, 2001 Orlando, Florida, vol 502388. ACM Press, pp 209–218

Habraken NJ (1972) Supports: an alternative to mass housing. Architectural Press, London

Hachimura K, Kato H, Tamura H (2004) A prototype dance training support system with motion capture and mixed reality technologies. In: 13th IEEE international workshop on robot and human interactive communication, ROMAN 2004, 20–22 Sept 2004, pp 217–222

Herman IP (2007) Physics of the human body. Springer, Berlin, New York

Hong T, Sun H, Chen Y, Taylor-Lange SC, Yan D (2016) An occupant behavior modeling tool for co-simulation. Energy Build 117:272–281

Hubers JC (2004) Muscle tower ii. TU Delft, Delft, The Netherlands. http://www.bk.tudelft.nl/ind ex.php?id=16060&L=1. Accessed 1 Nov 2013

Ishii H, Ullmer B (1997) Tangible bits: towards seamless interfaces between people, bits and atoms. In: CHI 1997. ACM Press, Atlanta, USA, pp 234–241

Jacobs M, Findley J (2001) Breathe. http://www.sonicribbon.com/sonicribbon/breathe/. Accessed 03 Mar 2015

Kohlstedt K (2004) Mirrored street facade art turns pedestrians into acrobats. http://weburbanis t.com/2013/01/27/mirrored-street-facade-art-turns-pedestrians-into-acrobats/. Accessed 10 Mar 2016

Kronenburg R (2002) Houses in motion: the genesis, history and development of the portable building. Academy Editions, London

Kronenburg R (2007) Flexible: architecture that responds to change. Laurence King, London

Kyan M, Sun G, Li H, Zhong L, Muneesawang P, Dong N, Elder B, Guan L (2015) An approach to ballet dance training through MS Kinect and visualization in a CAVE virtual reality environment. ACM Trans Intell Syst Technol 6:1–37

Larssen AT, Robertson T, Loke L, Edwards J (2007) Introduction to the special issue on movement-based interaction. Person Ubiquitous Comput 11:607–608

Latour B, Yaneva A (2008) Give me a gun and I will make all buildings move: an ANT's view of architecture. Explor Archit Teach Design Res:80–89

Lefebvre H (2013) Rhythmanalysis: space, time and everyday life. Bloomsbury Publishing PLC, London, UK

Loke L, Robertson T (2013) Moving and making strange: an embodied approach to movement-based interaction design. ACM Trans Comput-Hum Interact (TOCHI) 20:7

Long R (1967) A line made by walking, UK

Marshall J, Rowland D, Egglestone SR, Benford S, Walker B, Mcauley D (2011) Breath control of amusement rides. In: Proceedings of the sigchi conference on human factors in computing systems. ACM, Vancouver, BC, Canada

Montgomery GT (1994) Slowed respiration training. Appl Psychophysiol Biofeedback 19:211–225

Nehaniv CL, Dautenhahn K (2002) The correspondence problem. In: Kerstin D, Chrystopher LN (eds) Imitation in animals and artifacts. MIT Press

Phidgets INC (2006) Phidgets INC: unique and easy to use usb interfaces. Phidgets INC, Calgary, Canada. http://www.phidets.com. Accessed 12 June 2006

Processing Foundation (2016) Processing.Org. http://www.processing.org. Accessed 10 Mar 2016

Ricardo Cruz M, Nadia, B-B, Pablo R, Gustavo Casillas L (2015) A classification of user experience frameworks for movement-based interaction design. Design J 18:393–420

Rogers Y (2009) The changing face of human-computer interaction in the age of ubiquitous computing. In: Holzinger A, Miesenberger K (eds) 2009 Proceedings of the HCI and Usability for e-Inclusion: 5th symposium of the workgroup human-computer interaction and usability engineering of the austrian computer society, USAB 2009, Linz, Austria, November 9–10. Springer, Berlin, Heidelberg

Schnädelbach H (2010) Adaptive architecture-a conceptual framework. In: Geelhaar J, Eckardt F, Rudolf B, Zierold S, Markert M (eds) MediaCity. Bauhaus-Universität Weimar, Weimar, Germany

Schnädelbach H (2012) Hybrid spatial topologies. J Space Syntax 3:204–222

Schnädelbach H (2016) Movement in adaptive architecture. In: Griffiths S, Lünen AV (eds) Spatial cultures: towards a new social morphology of cities past and present. Routledge/Ashgate, London

Schnädelbach H, Glover K, Irune A (2010) ExoBuilding-breathing life into architecture. In: NordiCHI. ACM Press, Reykjavik, Iceland

Schnädelbach H, Irune A, Kirk D, Glover K, Brundell P (2012) ExoBuilding: physiologically driven adaptive architecture. ACM Trans Comput Hum Interact (TOCHI) 19:1–22

Schumacher M, Schaeffer O, Vogt, M-M (2010) MOVE: architecture in motion-dynamic components and elements, Birkhäuser

Solodkin A, Hlustik P, Buccino G (2007) The anatomy and physiology of the motor system in humans. handbook of psychophysiology. Cambridge University Press, Cambridge, UK

Studio Gang Architects (2009) Bengt sjostrom starlight theatre/studio gang architects. ArchDaily. http://www.archdaily.com/?p=28649. Accessed 02 Mar 2015

Thomsen MR (2008) Robotic membranes-exploring a textile architecture of behaviour. Protoarchitecture-analogue and digital hybrids. Wiley, London
Varela FJ, Thompson E, Rosch E (1991) The embodied mind: cognitive science and human experience. MIT Press, Cambridge
Weiser M (1991) The computer for the twenty-first century. Scient Am 265:94–104
Wheeler M (2005) Reconstructing the cognitive world. MIT Press, Cambridge, USA

Chapter 8
Designing and Prototyping Adaptive Structures—An Energy-Based Approach Beyond Lightweight Design

Gennaro Senatore (iD)

Abstract This chapter presents an overview of an original methodology to design optimum adaptive structures with minimum whole-life energy. Structural adaptation is here understood as a simultaneous change of the shape and internal load-path (i.e. internal forces). The whole-life energy of the structure comprises an embodied part in the material and an operational part for structural adaptation. Instead of using more material to cope with the effect of rare but strong loading events, a strategically integrated actuation system redirects the internal load path to homogenise the stresses and to keep deflections within limits by changing the shape of the structure. This method has been used to design planar and spatial reticular structures of complex layout. Simulations show that the adaptive solution can save significant amount of the whole-life energy compared to weight-optimised passive structures. A tower supported by an exo-skeleton structural system is taken as a case study showing the potential for application of this design method to architectural buildings featuring high slenderness (e.g. long span and high-rise structures). The methodology has been successfully tested on a prototype adaptive structure whose main features are described in this chapter. Experimental tests confirmed the feasibility of the design process when applied to a real structure and that up to 70% of the whole-life energy can be saved compared to equivalent passive structures.

8.1 Introduction

8.1.1 Context and Motivation

Civil structures (e.g. towers, bridges, stadia) are usually over-engineered for most of their service lives, as a result of being designed to withstand rare, worst-case loading

G. Senatore (✉)
Applied Computing and Mechanics Laboratory (IMAC), School of Architecture,
Civil and Environmental Engineering (ENAC), Swiss Federal Institute
of Technology (EPFL), CH-1015 Lausanne, Switzerland
e-mail: gennaro.senatore@epfl.ch

© Springer International Publishing AG, part of Springer Nature 2018
H. Bier (ed.), *Robotic Building*, Springer Series in Adaptive Environments,
https://doi.org/10.1007/978-3-319-70866-9_8

scenarios. Most of the time structures experience loads significantly lower than the design load and thus this requirement not only creates significant material wastage but it also restrains structural and architectural design.

Reducing the environmental impact of structures is now a serious concern in the construction industry. In a world going through critical changes due to energy depletion and financial challenges, there is a need for technologies and design methods that will transform the way we think about buildings in a way that is fit for purpose in the 21st century—which means lean, low-carbon and smart. This fact leads to consider that buildings could be adaptive rather than relying only on passive load-bearing capacity.

Adaptive structures are defined here as structures capable of counteracting the effect of loads via controlled shape changes and redirection of the internal load-path. In this context, structural adaptation means responding to external agents (e.g. mechanical and thermal loads) to keep the system within desired boundaries maintaining optimal performances throughout its service life. The main components of an adaptive structural system comprises: (1) sensing, (2) actuation, (3) control strategy (4) load bearing capacity (Yao 1972).

Sensing enables monitoring the system to gather data regarding its state. The state is made of key parameters belonging to mechanical and thermal physical domain (e.g. stress, strain, temperature) which are mapped as a function of space and time.

Actuation can be regarded as a controlled release of energy to keep the state of the system within desired boundaries. Actuation involves the transformation of stored (e.g. chemical) or supplied (e.g. electrical, magnetic) energy into mechanical energy. This energy can be utilised for example to control the shape of the structure or to varying its stiffness.

Information gathered by sensors are processed by a suitable **control strategy** to provide input commands to the actuators. For example, a feed-forward and feedback strategy where open and closed control loop are used simultaneously. The main difference between open and closed control strategies is that in the former the response of the system is not measured (Dorf and Bishop 2011). The open-loop system uses a mathematical model of the structural behaviour to predict control actions upon detecting external loads. In the closed-loop control instead, the response of the structure (e.g. stress, displacements, accelerations) to both external events and control actions is measured so that corrections can be made to achieve the desired state. Monitoring the structural response is important to achieve stable control due to the inherent inaccuracy of the process model and the impossibility to monitor all external disturbances.

Load-bearing capacity is achieved through the network of actuators and passive structural components arranged in space to form the structure to withstand static as well as dynamic loads. The main difference from a conventional passive structure is that some of elements of the system are active (i.e. the actuators) providing controlled output energy to manipulate the internal flow of forces and the shape rather than relying only on passive resistance (i.e. geometry and material).

A classification of structures with adaptive capabilities inspired by Wada et al. (1990) is illustrated in Fig. 8.1. Kinetic structures (Zuk and Clark 1970) are integrated

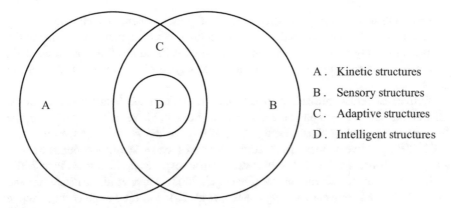

A . Kinetic structures

B . Sensory structures

C . Adaptive structures

D . Intelligent structures

Fig. 8.1 Adaptive structures classification

with actuators to perform motion e.g. a change of shape or position. Sensory structures are integrated with sensors in order to monitor the response to loading events. The intersection of kinetic and sensory structures are adaptive structures. Modern control strategies (e.g. machine learning, adaptive control) enable structures to learn from experience (i.e. response feedback stored in memory) to improve or change control laws over time to better cope with changing environments. Adaptive structures with learning capabilities can be thought of as "intelligent" structures which can anticipate response to external changes rather than only reacting to them (Shea and Smith 1998; Domer and Smith 2005).

8.1.2 Adaptation in Structural Applications

Active control of civil structures has focused mostly on the control of vibrations for building or bridges to improve on safety and serviceability during exceptionally high loads (e.g. strong winds, earthquakes) (Soong 1988). Active brace systems have been tested using hydraulic actuators fitted as cross-bracing elements of the structure controlling directly its deflections (Abdel-Rohman and Leipholz 1983; Reinhorn et al. 1992; Bani-Hani and Ghaboussi 1998). Deflection reduction in cable stayed bridges can be obtained via control forces provided by the stay cables working as active tendons (Rodellar et al. 2002; Xu et al. 2003). Active cable-tendons have been used to change the amount of pre-stress in reinforced concrete beams and in steel trusses to limit displacements under loading (Schnellenbach and Steiner 2013). Integration of actuators has been shown to be an effective way to suppress vibrations in high stiffness-to-weight ratio truss structures (Preumont et al. 2008).

Actuation has been used to modify the membrane stress state in thin plates and shells to help them cope with unusual loading events (Weilandt 2007). Residual stresses formed after welding, machining or formworks removal (Sobek 1987) can reduce shells load carrying capacity significantly. In the event of such disturbances, actuation in the form of induced strain distributions or induced displacements of the supports (actively controlled bearings) can be used to homogenise the stress field and in so doing minimizing the maximum stress governing the design (Neuhäuser 2014).

Active structural control has also been used in applications for shape control. Some all-weather stadia use deployable systems for expandable/retractable roofs e.g. the Singapore National Stadium (Henry et al. 2016) and the Wimbledon Centre Court (SCX 2010). Active tensegrity structures, structures whose stability depends on self-stress, have been used for deployable systems in aerospace applications (Tibert 2002) as well as for displacement control (Fest et al. 2003; Veuve et al. 2015; Adam and Smith 2008) and frequency tuning (Santos et al. 2015) in civil engineering. Active compliant structures, which can be thought of as structures working as monolithic mechanisms (Campanile 2005; Hasse and Campanile 2009), have been investigated for the deployment of antenna reflectors (Jenkins 2005), for the control of aircraft wings to improve on manoeuvrability (Previtali and Ermanni 2012) as well as for the control of direct daylight in buildings (Lienhard et al. 2011).

Because of uncertainties regarding the long-term reliability of sensor and actuator technologies combined with building long service lives and load long return periods, the recent trend has been to develop active structural control to help satisfy serviceability requirements (e.g. deflection limits) rather than contribute to strength improvement (Korkmaz 2011). If the structure relies on an active system for deflection control, its stiffness can be distributed strategically to better utilise the material.

In this context, the relevance of adaptive structures is significant. Advances in material science have mainly focused on increasing the strength of commonly used materials such as steel and concrete but not their stiffness thus leading to problems to satisfy serviceability requirements (Connor 2002). The trend to build slender, taller and longer span structures is shifting design criteria from strength to serviceability where motion control (i.e. limitations of displacements and accelerations) is one of the main issues. In addition, there has been an increase in the use of special structures for space applications, manufacturing and transport facilities that must meet strict design constraints for serviceability (Connor 2002).

8.2   Optimum Design Methodology for Adaptive Structures

In the natural world living forms and their structure are optimised around a strategic balance between material and energy resources (Vincent 1990). The metabolic cost of being adaptive to reach resources is carefully traded with the cost of energy embodied in the material. Most existing design strategies for adaptive structures are based on optimisation methods which aim at minimizing a combination of the control effort, structural response to external loads and other cost functions including the mass of the structure (Utku 1998; Teuffel 2004; Soong and Pitarresi 1987; Sobek and Teuffel 2001). However, whether the energy saved by using less material makes up for the energy consumed through control and actuation is a question that has so far received little attention.

A new optimum design methodology for adaptive structures was presented in Senatore et al. (2011, 2013). This method is based on improving building structural performances by reducing the energy embodied in the material for extraction and manufacturing at the cost of a small increase in operational energy necessary for structural adaptation and sensing. The design process comprises two main steps: (1) embodied energy optimisation and (2) operational energy computation nested within an outer optimisation minimising the whole-life energy. Whole life energy is here understood as the sum of the embodied energy in the material and the operational energy used by the active control system. Applying this methodology to a range of planar and complex spatial reticular structures, Senatore et al. (2018a) show that the adaptive solution can achieve energy savings as high as 70% when compared to identical weight-optimised passive structures. These studies confirm that adaptive structures achieve superior performance when the design is stiffness governed.

The method has so far been implemented for reticular structures with statically determinate and indeterminate topologies. Figure 8.2 shows a schematic flowchart of the design process. Each step is illustrated on a truss structure case study which is employed here as a visual aid.

8.2.1 Inputs

Inputs include the structural topology, material and type of elements, loading and deflection limits (serviceability limit state). In this case, the input layout is a catenary structure which could be thought of as a section of a truss arch bridge made of steel tubular elements. The supports are all pinned as indicated in Fig. 8.2a. The structure is subjected to a uniformly distributed dead load and two patch loads each covering half span of the bridge (e.g. vehicular traffic or train holding position). Serviceability limits are set to be a fraction of the span (e.g. span/1000 typically used for road bridges with both vehicular and pedestrian traffic (Barker et al. 2011).

Inputs include the selection of certain parts of the structure which are of critical importance to serviceability and therefore should be controlled by the active system. For example, in this case all the nodes of top chord except the supports are selected to be controlled as indicated by the circles in Fig. 8.2a. Finally, a suitable range must be chosen for the material utilisation factor (MUT). This MUT is a ratio of the strength capacity over demand but it is defined for the structure as a whole and can be effectively thought of as a scaling factor on the allowable stresses. The MUT varies in a range of $0\% < \text{MUT} \leq 100\%$.

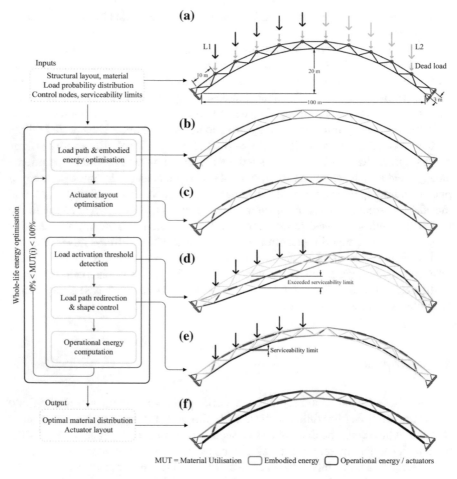

Fig. 8.2 **a** Initial layout and controlled nodes; **b** minimum embodied energy design MUT = 100%; **c** optimal actuation layout MUT = 100%; **d** deformed shape; **e** controlled shape; **f** minimum whole-life energy design MUT < 100%

8.2.2 Load-Path and Embodied Energy Optimisation

The embodied energy of the structure is minimised by optimising the internal load paths and corresponding material distribution but ignoring serviceability limit states, thus obtaining a lower bound in terms of material mass. The energy analysis is carried out using a factor to convert material mass into embodied energy (Hammond and Jones 2008). The design variables are element cross-section areas and internal forces satisfying equilibrium and "strength" constraints (admissible stress, element instability) when the structure is subjected to the design loads. This problem has been solved using sequential quadratic programming (Senatore et al. 2013).

Strength constrains (ultimate limit states, ULS) include the MUT which factors the material yield stress in tension and compression. This way, by varying the MUT one can move from least-weight structures (MUT = 100%) with small embodied but large operational energy, to stiffer structures with large embodied and smaller operational energy consumption. At this stage, the MUT is constant between 0 and 100%. Figure 8.2b shows the configuration obtained for an MUT of 100% which corresponds to the absolute minimum embodied energy structure but it might result in a high level of operational energy for structural adaptation. The MUT is the main variable of the embodied-operational energy optimization (Sect. 8.2.5) to obtain an optimum compromise between active and passive design.

The internal force vector (i.e. load-paths) is called optimal or non-compatible. This is because, at this stage, deflection constraints (serviceability limit state, SLS) as well as geometric compatibility constraints (i.e. all elements connected to a node must have the same absolute displacement) are intentionally not included. When external loads are applied to the structure, the compatible forces will, in general, be different from the optimal forces and the resulting displacements might be beyond serviceability limits. For this reason, the next step is to design the actuation system which involves to determine the location of the actuators. The actuators are mechanical devices (e.g. linear motors) which are thought of as integrated into the structure by replacing some of its members. The actuators produce work in the form of length changes. The effect of such length changes is to manipulate the internal forces to match the optimal load-path (i.e. enforce geometric compatibility) and to reduce deflections within required serviceability limits by changing the shape of the structure.

8.2.3 Actuator Layout Optimisation

For a discretised structure (e.g. a truss) the actuator placement problem is of combinatorial nature as it involves selecting a certain number of actuator locations from a set of available sites (the structural elements). This type of problem is usually solved employing global search methods (e.g. stochastic) which can become computationally very expensive and impractical for structures made of many elements. However, in this work the actuator placement problem is formulated as a least-square constrained optimisation via a sensitivity analysis. For each element in turn, the efficacy to redirect the load-path and to correct displacements of a unitary change in length is assessed. Then the difference between the nodal displacements caused by the element length changes and the required displacement correction subject to geometric compatibility constraints is minimised.

This process produces a ranking indicating how effective each element of the structure would be if it was replaced by an active element. This way the most effective locations are selected to form the actuator layout. The minimum number of actuators to control the required displacements exactly is equal to the sum of the number of assigned controlled degrees of freedom and the static indeterminacy of the structural system. Intuitively this is the number of actuators needed to turn the structure into a controlled mechanism. One actuator can control at least one degree of freedom and for statically indeterminate structures extra actuators, as many as the number of static indeterminacy, are needed to control the internal forces. In case fewer actuators are fitted into the structure, displacements can still be compensated albeit only approximately.

8.2.4   Operational Energy Computation

8.2.4.1   Load Probability Distribution

The computation of the operational energy requires assuming some statistics on the frequency of occurrence of the loads. It is intuitively clear that the proposed design process can be particularly beneficial when the design is governed by large loading events having a small probability of occurrence such as storms, earthquakes, snow, unusual crowds etc. Probabilistic models already exist for most of these loads. For instance, earthquakes are often modelled with a Poisson distribution and wind storms with a Weibull distribution (Flori and Delpech 2010). Should this methodology be applied in a practical case, the relevant load probability distribution should be used.

For the purpose of describing the design methodology in this chapter, it is more convenient to work with a generic distribution which can be easily parametrised to fit different loading scenarios. The load probability distribution is modelled here using a Log-Normal distribution because it is closely related to the Normal probability distribution, hence it is general, only taking positive real values and thus providing the desired bias toward the lower values of the random variable. For simplicity, the mean of the underlying normal distribution is set to zero. Following the limit-state design methodology, the characteristic load (i.e. the design load) is set as the 95th percentile of the probability distribution (Nowak and Collins 2012). Once the mean and the characteristic load are set, the standard deviation is fully determined. The design life is usually set to 50 years. The effect of the assumptions made here regarding the load probability distribution are tested systematically in (Senatore et al. 2018b).

8.2.4.2   Load Activation Threshold

The hybrid passive-active structural system is designed so that in normal loading conditions it can take the load using only its passive capacity with the actuators locked in position. The actuators are only activated when the loads reach an activation threshold which is the load causing a state of stress violating either an ultimate (ULS) or a serviceability limit state (SLS). This means that only the rarer loads with higher magnitude but less probability of occurrence need both passive and active load-bearing capacity and therefore the operational energy will be only used when necessary. The introduction of the load activation threshold shows how passive and active design can be combined to reach a higher level of efficiency. Passive resistance through material and form is replaced by a small amount of operational energy.

8.2.4.3 Load Path Redirection, Shape Control and Actuator Work

For any load above the load activation threshold the active system must redirect the internal load-path and control the shape of the structure. For instance, Fig. 8.2d shows that the nodal displacements caused by the design load exceed serviceability limits. Figure 8.2e shows the shape controlled via actuation whereby all controlled displacements are reduced to the required limits.

To compute the operational energy, further assumptions have to be made regarding the mechanical efficiency and the working frequency of the actuators. The mechanical efficiency depends on the actuation technology. For instance, hydraulic actuators have a mechanical efficiency in a range 90–98% (Huber et al. 1997). To be conservative, it is assumed that actuation is hydraulic with a mechanical efficiency of 80%. The mass of an actuator is assumed to be a linear function of the required force with a constant of 0.1 kg/kN (ENERPAC 2016).

It is assumed that the actuators always work at the first natural frequency of the structure which is likely to dominate the response of most structures excited by dynamic loads relevant to civil engineering. This assumption is conservative because it implies that even if the loads only vary very slowly in time, the actuators work at the first natural frequency of the structure. It is also assumed that non-active means are employed to control vibrations (e.g. tuned mass dampers) if required.

During structural adaptation, each actuator does work to change length under resisting forces. The sum of all actuators work divided by the mechanical efficiency and multiplied by the working frequency times the hours of occurrence of a load, is the energy spent for a particular load occurrence above the activation threshold. Summing over all loads above the activation threshold gives the energy needed for structural adaptation throughout the structure service life.

8.2.5 Minimum Whole-Life Energy Design

The outer optimisation performs a search for the optimal Material Utilisation Factor (MUT). For each MUT, the embodied energy and internal load-paths are optimised (Sect. 8.2.2), subsequently the optimal actuator layout is obtained (Sect. 8.2.3) and the operational energy is computed (Sect. 8.2.4). Figure 8.3 shows notionally the variation of the embodied and operational energy as well as their sum (i.e. total or whole-life energy) as the MUT varies. The active-passive system corresponding to the minimum of whole-life energy is the configuration of the optimum sought. Figure 8.2f shows the minimum whole-life energy design for the case study defined in Sect. 8.2.1. Although this structure has a higher embodied energy than that obtained for an MUT of 100% shown in Fig. 8.2c, its whole-life energy is lower because it requires a much lower operational energy for structural adaptation.

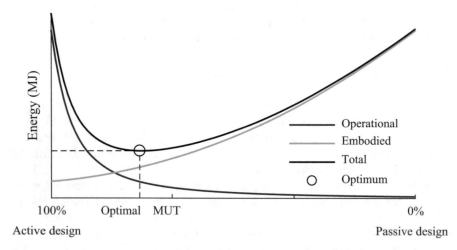

Fig. 8.3 Embodied, operational and total energy as a function of the material utilisation factor (MUT) © IOP Publishing. Reproduced with permission. All rights reserved (Senatore et al. 2018c)

The comparison between Fig. 8.2c and f shows that the embodied energy optimisation (Sect. 8.2.2) and the actuator layout optimisation (Sect. 8.2.3) are very much interdependent. This is because the actuators, by changing the shape of the structure to meet deflection requirements, allow it to be much leaner with lower embodied energy. Conversely, the actuator optimal layout is dependent on the structure within which the actuators are to be fitted. The efficacy of an actuator regarding force and displacement control depends, aside from its location in the structure and the position of the controlled nodes, from the contribution of the passive load-bearing capacity. When varying the MUT, the resulting material distribution changes thus requiring a different internal load-path redirection and displacement compensation. For this reason, the actuator optimal layout changes for different values of the MUT.

8.2.6 Structural Adaptation Simulation

Structural adaptation is here understood as a controlled change of the shape and internal forces of the structure. Simulating a controlled shape change is not a trivial task. For truss systems such as those described in Sects. 8.2 and 8.3, a shape change is the result of simultaneous expansion or contraction of the actuators that are fitted into the structure. Even for a truss system (which is made of elements that can only be either in tension or compression) most of commercially available simulation tools do not offer a way to assign directly an extension or contraction of one or more elements in the structure.

A convenient method to compute internal forces and displacements resulting from element length changes is the Integrated Force Method (IFM) (Patnaik 1973). The IFM was originally formulated to allow a geometric imperfection, caused by a lack of fit or thermal strains for instance, to be dealt with in a compact way and without the need to choose any specific member as redundant. This is because in statically indeterminate structures, internal stresses can be caused by geometrical imperfections and therefore should be taken into account in structural design. In the design method outlined in Sect. 8.2, a deformation vector akin to a lack of fit is defined to assign the actuator length changes. The length change of an actuator is thought of as a non-elastic strain which is referred as eigenstrain (Ziegler 2005). Shape control and internal load path redirection simulation (see Sect. 8.4.3) is handled by a computationally efficient routine based on eigenstrain assignment via the Integrated Force Method. This routine is described comprehensively in Senatore (2016).

An alternative way to simulate controlled shape changes via actuation is given in Senatore and Piker (2015) which presented a formulation combining the principle aspects of the Dynamic Relaxation method (Day 1965; Barnes 1977; Williams 2000) and the co-rotational formulation for the Finite Element Method (Crisfield 1990; Felippa and Haugen 2005). In this formulation, elements forces, moments and inertia are appropriately lumped at nodes. Position, velocity and acceleration of each node are computed iteratively. A co-rotational approach is employed to compute the resultant field of displacements in global coordinates including the effect of large deformations (i.e. geometric non-linearity). The system converges to an equilibrium position around which it oscillates and eventually settles when the out of balance forces and moments residuals are below a set tolerance. This formulation was implemented into a cross-platform software called "PushMePullMe" (Senatore 2017a) written in Java and later as the game "Make A Scape" (Senatore 2017b) running on the mobile operating system iOS.

Since convergence to equilibrium is iterative, it is possible to change interactively the length of an element (often referred as "rest-length" in this context) to simulate its expansion or contraction. In addition, because computation of displacements and forces relies only on the local element stiffness matrix (i.e. there is no need to assemble a global stiffness matrix), changing the length of an element does not take substantial computational resources. For illustration purposes, Fig. 8.4 shows (a) an hypothetical roof truss structure whose top chord elements are replaced by actuators (indicated in purple) and (b) the shape change obtained by reducing the length of all the actuators by 10% the initial length.

(a)

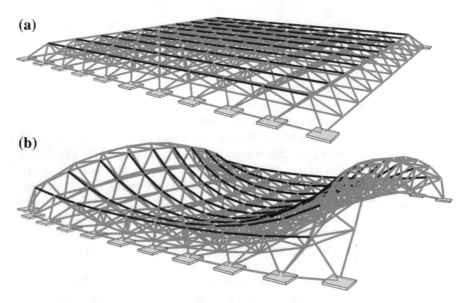

(b)

Fig. 8.4 Adaptive roof structure: **a** initial geometry; **b** shape change after 10% actuator length reduction

8.3 Case Study

The structure considered in this case study is a simplified model of a tower building known as 30 St Mary Axe or informally the "Gherkin", a tall building in the City of London. This case is part of a parametric study that has been carried out to investigate how adaptive structure performances in terms of mass and energy savings as well as monetary costs vary when the design process is applied to complex spatial configurations (Senatore et al. 2018a). The model is loosely related to the original geometry which is studied here as an example of a tall building resisting external loads through an exoskeleton structure. This means that the example studied here has no structural core (although the real 'Gherkin' does). As cores reduce significantly commercially usable floor space, systems that can do away with them free up the floor layouts and are therefore of structural, architectural and commercial interest.

Two models are considered to show how energy savings vary with the slenderness i.e. the ratio height to depth (H/D). Main dimensions and boundary conditions are indicated in Fig. 8.5. All elements are assumed to have a cylindrical hollow section. To limit the complexity of the optimisation process, the element wall thickness is set to 10% of the external diameter. Limit to the total building drift is set to height/500. The horizontal displacements of all the nodes except the supports are controlled.

Five load cases are considered. L1 is self-weight+dead load which is set to 3 kN/m^2 on the floors of the building and transmitted on the nodes of the exoskeleton structure. The live load consists of four wind-type load cases arranged in two pairs with opposite directions. Figure 8.5c shows a top view of the structure with (c) L2 (symmetrical to L4) and (d) L3 (symmetrical to L5) applied. The live load intensity

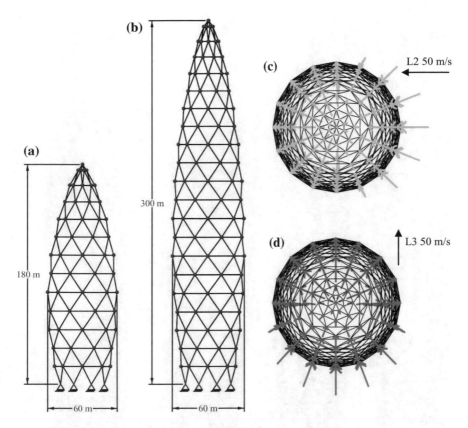

Fig. 8.5 Dimensions and control nodes indicated by dots **a** H/D=3. **b** H/D=5; **c** L2; **d** L3. Reproduced with permission from ASCE (Senatore et al. 2018a)

varies quadratically with the height reaching a maximum of 1.5 kN/m^2 which is equivalent to a wind velocity of 50 m/s (category 2/3 hurricane). All live load cases have identical probability distribution (see Sect. 8.4.1).

The activation thresholds are 1.0 kN/m^2 and 0.7 kN/m^2 when the H/D ratio is 3 and 5 respectively. In terms of wind velocity, the activation thresholds correspond to approximately 40 and 34 m/s and the total time during which actuation is required to compensate for deflections is 1.25 and 3 years. Mass and total energy savings compared to a passive structure of identical layout designed using a state of the art optimisation method (Patnaik et al. 1998) are 25 and 8% for H/D=3 and 48 and 31% for H/D=5. The optimal adaptive structure is obtained for an MUT of 51% for H/D=3 and 43% for H/D=5. This is because for a higher H/D (i.e. a more slender structure), displacement compensation takes more operational energy and thus the MUT decreases.

Figure 8.6 compares the passive structure (a) with the adaptive structure (b) for the case H/D=5. As expected the actuator layout (represented in purple) is denser towards the bottom of the structure because it is the most effective location for

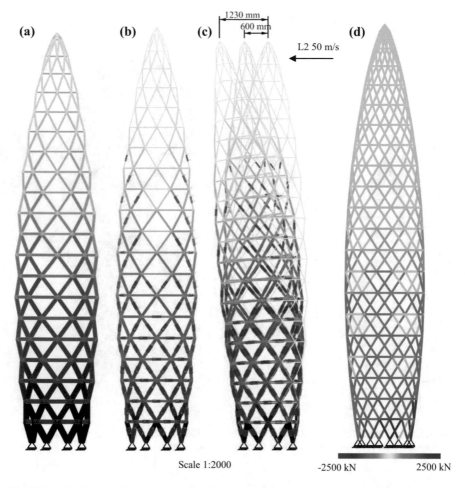

(a)   **(b)**   **(c)**   1230 mm   600 mm   L2 50 m/s   **(d)**

Scale 1:2000

-2500 kN   2500 kN

Fig. 8.6 **a** Passive, **b** adaptive, **c** controlled & deformed shape (mag. 50 ×), **d** load path redirection
Reproduced with permission from ASCE (Senatore et al. 2018a)

the actuator length changes to reduce the top nodes displacements. Without active
displacement compensation, the maximum deflection is beyond serviceability limit
(height/500 = 600 mm) as shown in Fig. 8.6c. The load path redirection (difference
between optimal and compatible forces) for L2 is illustrated in Fig. 8.6d. Matching
the optimal load path requires adding compressive forces on the side the wind load
hits the structure and on the opposite side which is subjected to negative pressure.
The maximum length changes are about 40 mm expansion made by the bottommost
actuators located on the opposite side the load is applied which must deploy under the
highest compressive forces (35,000 kN). The highest tensile forces are approximately
18,000 kN applied by the actuators placed on the horizontal elements (it can be
thought of as the action of tightening "rings" on a basket-like structure).

8.4 Experimental Prototype

A large scale prototype (here named "adaptive truss"), designed using the method outlined in Sect. 8.2, was built at University College London Structures Laboratory. The prototype is a 6 m cantilever spatial truss with a 37.5:1 span-to-depth ratio consisting of 45 passive steel members and 10 electric linear actuators strategically fitted within the tension diagonal members. The adaptive truss main dimensions are shown in Fig. 8.7.

The truss is designed to support its own weight which consists of 52 kg for the steel structure, 50 kg for the actuators (5 kg each) and 70 kg for the acrylic deck panels and housing. The live load is thought of as a person walking along the deck. This is modelled as three load cases representing the worst scenarios when the person stands at the free end with their weight distributed equally between the two end nodes or when their entire weight is concentrated on either one of them. The magnitude of the live load is set to 1 kN (100 kg). Deflection limits are set to span/500 (12 mm) because due to its pronounced slenderness, this truss can be regarded as the scaled super structure of a tall tower subjected to wind load. The members of the structure are sized to meet the worst expected 'demand' from all load cases to be fully compliant to Eurocode 3 in terms of ultimate limit state but ignoring deflection requirements.

The deck/façade of the structure consists of a series of aluminium angle profiles which house transparent acrylic panels Fig. 8.8a. Clear acrylic has been chosen to allow the actuator length changes to be seen during control. The aluminium angles also provide housing to power and signal cables which are bundled and clipped to their bottom face as shown in Fig. 8.8b.

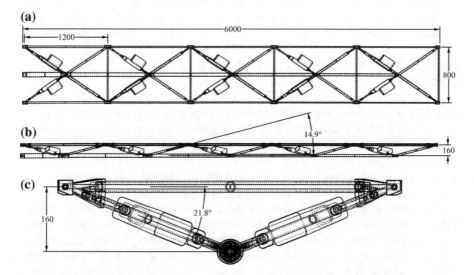

Fig. 8.7 Adaptive truss dimensions, **a** plan view, **b** elevation, **c** side view © IOP Publishing. Reproduced with permission. All rights reserved (Senatore et al. 2018c)

Fig. 8.8 **a** Deck/façade; **b** signal and power cables clipped underneath the aluminium angles ©
IOP Publishing. Reproduced with permission. All rights reserved (Senatore et al. 2018c)

The control system architecture has been designed with the primary aim to achieve identification of the response to loading in terms of internal forces and displacements for the structure to be able to be control itself without user intervention. The control system comprises ten linear actuators, a control driver board for each pair of actuators, 45 strain gauge based sensors, two amplifiers for signal conditioning and a main controller for acquisition and processing. The deformation of each element together with the actuator stroke position feedback are fed into the main controller. These inputs are processed to reconstruct the node spatial positions to assess whether their displacements exceed the required serviceability limit. In this case, the actuators vary their stroke length to change the shape of the structure so that controlled node displacements are reduced within required limits. To keep power consumption to a minimum, a switch on/off command is sent to the actuator driver to cut off power supply as soon as the target position is reached. The reader is referred to (Senatore et al. 2018c) for a detailed presentation of the prototype including the control algorithm.

Extensive loads tests has shown that the displacements are practically reduced to zero with no prior knowledge of direction, position and magnitude (within limits) of the external load thus achieving an "infinite" stiffness structure (i.e. zero deflection under loading). When a person walks on the deck, the actuators change length continuously to compensate for displacements as the load changes position. Figure 8.9 shows an example of the difference between the deformed (i.e. without control) and controlled shape. Demonstration movies are available online (Senatore 2017c).

Current sensors were installed at the mains supply to monitor the power used for shape control. Energy consumption was recorded during displacement compensation under quasi-static loading for all electronic devices including the actuation system, signal conditioning and main control processor. The live load probability distribution (Sect. 8.4.1) was divided in 10 steps from 10 to 100 kg. Each load (in the form of weights) was placed on the deck between the end nodes. The total operational energy

Fig. 8.9 Person walking (70 kg), comparison between deformed (transparent) and controlled shape © IOP Publishing. Reproduced with permission. All rights reserved (Senatore et al. 2018c)

was computed by multiplying the energy consumption needed for shape control by the live load hourly distribution for all loads above the activation threshold (a weight of 14 kg at the free end). The adaptive truss prototype has been benchmarked against two passive structures designed to cope with identical loads and deflection limits. The first structure is made of two steel I-beams while the second is an equivalent truss designed using a state-of-the-art optimisation method (Patnaik et al. 1998). Measurements has shown that the adaptive truss achieves 70% energy savings compared to the I-beams and 40% compared to the optimised passive truss.

Using fast-acting actuators would allow full control of dynamics/vibrations as well as deflections. The actuators used in this prototype were readily available from the automotive industry (at relatively low cost) and move at 11 mm/s. Nevertheless, they are still able to increase effective damping in the truss from 0.5% to 3%. However, actively controlling vibrations expends much more operational energy therefore hybrid solutions using actuation to compensate for large displacements and passive means (e.g. tuned mass dampers) to control vibrations are still likely to be preferred.

This prototype was also built as a demonstration piece to show in a practical and interactive way the potential of the underlying design methodology to professionals in the field—structural engineers, architects, fabricators. The structure was exhibited at various key institutions amongst with University College London and during the International Association for Shell and Spatial Structures symposium (IASS) held in Amsterdam in 2015. A month solo exhibition took place in August 2016 at a well-known building technology gallery space called "The Building Centre" situated in central London (Senatore 2017d).

8.5 Discussion

Adaptive structures offer an emerging design paradigm that deals with providing stiffness in a completely different way to traditional engineering. Using adaptation as a strategy to counteract the effect of the external load allows large quantities of materials to be saved while meeting safety critical requirements:

(1) Conventional materials e.g. steel tubes or rods as shown in the prototype presented in this chapter, still provide for strength and safety (ultimate limit state requirements) as well as for deflections under day-to-day loads;
(2) The actuation system controls excessive movements and deflections (serviceability limit state) which in practice occur very infrequently;
(3) In case of a power outage the actuators simply stop moving (i.e. fail-safe) but load carrying capacity is not compromised because of point 1.

The design method proposed in this work produces structures that combine three objectives which are usually mutually exclusive: (1) the structure has a low overall environmental impact (minimum whole-life energy design); (2) the displacements can be controlled within very tight limits; (3) the structure is extremely slender. Being able to combine these three objectives is unique in structural engineering and architecture. In the case of the prototype structure described in Sect. 8.4 a combination of all three benefits has been in fact achieved.

Applying this design philosophy, scenarios where adaptive structures could bring significant benefits include:

– When the end use has very stringent/high performance requirements for deflection, therefore the "infinite stiffness" capability of adaptive structures can clearly outperform conventional structures. For example, laboratory buildings, gantry crane runway beams, bespoke facades etc.
– When the structural design is governed by high but rare loads, such as earthquakes and wind storms. The same applies to structures that are in service for only a few hours per week (e.g. stadium stands). An adaptive structure optimised to remain serviceable under rare high loads can give 80% material weight savings compared to conventional passive structures.
– Adaptive structures are technically well suited to architectural buildings where very high slenderness/shallow structural depths are needed. This could be either a pure aesthetic driver, limited floor-ceiling heights in a new building, or limited space for new structure in a complex refurbishment.
– Long span/high rise structures are typically stiffness governed and so tall buildings, bridges, and roofs could all see benefit from adaptive design. These types of structures would likely take a combination of all three characteristics (stiffer, lower weight, more slender). For example, an architectural footbridge could be very slender and at the same time lighter than normal, in order to install over a railway in a single crane lift. Tall buildings could have a much smaller stability core and smaller footprint.

8.6   Conclusions

This chapter outlines an optimum methodology to design adaptive structures. Structural adaptation is employed to counteract the effect of loads. The novelty of this work lies in the development of a methodology that produces, given any stochastic occurrence distribution of the external load, an optimum design of the structure for minimum whole-life energy comprising an embodied part in the material and an operational part for adaptation. The case study showed that even for complex structures, significant energy savings can be achieved, the more so the more stiffness-governed the structure is. Experimental tests confirmed the feasibility and applicability of the design method and that for slender configurations adaptive structures achieve substantive total energy savings compared to passive structures.

The method proposed here works within the assumption of small displacements and was implemented for statically determinate and indeterminate reticular structures. Ongoing work (Reksowardojo et al. 2017; Senatore et al. 2017) is exploring large shape changes (i.e. finite displacements) to allow for a full utilisation of the shape adjustable properties of adaptive structures for the purpose of saving energy. Future work could extend this design method to other structural systems (e.g. beams, shells) and investigate structural adaptation using material with non-linear behaviour (e.g. concrete).

Acknowledgements The author gratefully acknowledges the Engineering and Physical Sciences Research Council (EPSRC) who provided core funding through UCL Doctoral Training Centre in Urban Sustainability and Resilience, Expedition Engineering who provided significant additional resources. EPFL Applied Computing and Mechanics Laboratory (IMAC) is thankfully acknowledged for their support during the review process of this chapter.

References

Abdel-Rohman M, Leipholz H (1983) Active control of tall buildings. J Struct Eng 109(3):628–645
Adam B, Smith IF (2008) Active tensegrity: a control framework for an adaptive civil-engineering structure. Comput Struct 86(23–24):2215–2223
Bani-Hani K, Ghaboussi J (1998) Nonlinear structural control using neural networks. J Eng Mech 124(3):319–327
Barker GM, Staebler J, Barth K (2011) Serviceability limits and economical steel bridge design (report no. FHWA-HIF-11-044). U.S. Department of Transportation, Washington, DC
Barnes MR (1977) Form finding and analysis of tension space structures by dynamic relaxation. Doctoral dissertation, City University, London
Campanile LF (2005) Initial thoughts on weight penalty effects in shape-adaptable systems. J Intell Mater Syst Struct 16:47–56
Connor JJ (2002) Introduction to structural motion control. Pearson Education, Boston
Crisfield M (1990) A consistent co-rotational formulation for non-linear, three-dimensional, beam-elements. Comput Methods Appl Mech Eng 81(2):131–150
Day A (1965) An introduction to dynamic relaxation. Engineer 220–221
Domer B, Smith I (2005) An active structure that learns. J Comput. Civ. Eng 19(1):16–24
Dorf RC, Bishop RH (2011) Modern control systems, 12th ed., Pearson

ENERPAC (2016) E328e industrial tools—Europe. http://www.enerpac.com/en-us/downloads. Accessed 12 July 2017

Felippa C, Haugen B (2005) A unified formulation of small-strain corotational finite elements: I. Theory. Comput Methods Appl Mech Eng 194(21–24):2285–2336

Fest E, Shea K, Domer B, Smith F (2003) Adjustable tensegrity structures. J Struct Eng 129:515–526

Flori JP, Delpech GG (2010) Stavros Niarchos foundation cultural center in Athens Part I: climatic analysis. Technical report, Centre Scientifique et Tecnique du Batiment, Nantes

Hammond G, Jones C (2008) Embodied energy and carbon in construction materials. In: Proceedings of the institution of civil engineers—energy, vol 161, no 2, pp 87–98

Hasse A, Campanile F (2009) Design of compliant mechanisms with selective compliance. Smart Mater Struct 18(11)

Henry A, Kam C, Smith M, Lewis C, King M, Boulter N, Hoad P, Wong R, Munro S and Ming S (2016) Singapore sports hub: engineering the national stadium. Struct Eng 94(9)

Huber JE, Fleck NA, Ashby MF (1997) The selection of mechanical actuators based on performance indices. In: Proceedings: mathematical physical and engineering sciences, vol 453, pp 2185–2205

Jenkins C (2005) Compliant structures in nature and engineering, 1st edn. WIT Press

Korkmaz S (2011) A review of active structural control: challenges for engineering informatics. Comput Struct 89:2113–2132

Lienhard J, Schleicher S, Poppinga S, Masselter T, Milwich M, Speck T, Knippers J (2011) Flectofin: a hingeless flapping mechanism inspired by nature. Bioinspir Biomimet 6:1–7

Neuhäuser S (2014) Untersuchungen zur Homogenisierung von Spannungsfeldern bei adaptiven Schalentragwerken mittels Auflagerverschiebung. University of Stuttgart (ILEK), Stuttgart

Nowak AS, Collins KR (2012) Reliability of structures, 2nd edn. Taylor & Francis

Patnaik S (1973) An integrated force method for discrete analysis. Int J Numer Meth Eng 6:237–251

Patnaik S, Gendy A, Berke S, Hopkins D (1998) Modified fully utilized design (MFUD) method for stress and displacement constraints. Int J Numer Meth Eng 41:1171–1194

Preumont A, de Marneffe B, Deraemaeker A, Bossensb F (2008) The damping of a truss structure with a piezoelectric transducer. Comput Struct 86(3–5):227–239

Previtali F, Ermanni P (2012) Performance of a non-tapered 3D morphing wing with integrated compliant ribs. J Smart Mater Struct 21:1–12

Reksowardojo AP, Senatore G, Smith IF (2017) Large and reversible shape changes as a strategy for structural adaptation. In: International associtaion for shell and spatial structures, Hamburg

Reinhorn AST, Lin R, Riley M (1992) Active bracing sytem: a full scale implementation of active control. National Center for Earthquake Engineering Research, Buffalo

Rodellar J, Mañosa V, Monroy C (2002) An active tendon control scheme for cable-stayed bridges with model uncertainties and seismic excitation. Struct Control Health Monit 9(1):75–94

Santos FA, Rodrigues A, Micheletti A (2015) Design and experimental testing of an adaptive shape-morphing tensegrity structure, with frequency self-tuning capabilities, using shape-memory alloys. Smart Mater Struct 24:1–10

Schnellenbach MH, Steiner D (2013) Self-tuning closed-loop fuzzy logic control algorithm for adaptive prestressed structures. Struct Eng Int 163–172

SCX (2010) Wimbledon centre court retractable roof. http://www.scxspecialprojects.co.uk/cache/f ilelibrary/73/library/fileLibrary/2011/6/Wimbledon.pdf. Accessed 15 Sept 2016

Senatore G, Duffour P, Hanna S, Labbe F, Winslow P (2011) Adaptive structures for whole life energy savings. Int Assoc Shell Spat Struct (IASS) 52(4):233–240

Senatore G, Duffour P, Winslow P, Hanna S, Wise C (2013) Designing adaptive structures for whole life energy savings. In: Proceedings of the fifth international conference on structural engineering, mechanics & computation, Cape Town. Taylor & Francis Group, London, pp 2105–2110

Senatore G, Piker D (2015) Interactive real-time physics: an intuitive approach to form-finding and structural analysis for design and education. Comput Aided Des 61:32–41

Senatore G, Duffour P, Winslow P (2018a) Energy and cost assessment of adaptive structures: Case studies. J Struct Eng (ASCE) 144(8):04018107

Senatore G, Duffour P, Winslow P (2018b) Exploring the domain of application of adaptive structures. Eng Struct 167:608–628

Senatore G, Duffour P, Winslow P, Wise C (2018c) Shape control and whole-life energy assessment of an "infinitely stiff" prototype adaptive structure. Smart Mater Struct 27(1):015022

Senatore G (2016) Adaptive building structures. Doctoral dissertation, University College London, London

Senatore G (2017a), PushMePullMe 3D. http://www.gennarosenatore.com/research/real-time_phy sics/push_me_pull_me_3d.html. Accessed 09 Nov 2017

Senatore G (2017b) Make a scape. http://www.gennarosenatore.com/projects/make_a_scape.html. Accessed 09 Nov 2017

Senatore G (2017c) Adaptive structures demonstration movies. https://vimeo.com/groups/adaptiv estructures. Accessed 03 2017

Senatore G (2017d) Adaptive structures—building centre exhibition. http://www.gennarosenatore. com/research/adaptive_structures/the_building_centre_exhibition.html. Accessed 19 Sept 2017

Senatore G, Wang Q, Bier H, Teuffel P (2017) The use of variable stiffness joints in adaptive structures. In: International association for shells and spatial structures, Hamburg

Shea K, Smith I (1998) Intelligent structures: a new direction in structural control, Berlin

Sobek W (1987) Auf pneumatisch gestützten Schalungen hergestellte Betonschalen. Doctoral dissertation, University of Stuttgart, Stuttgart

Sobek W, Teuffel P (2001) Adaptive systems in architecture and structural engineering. In: Liu SC (ed) Smart structures and materials 2001: smart systems for bridges, structures, and highways, Proceedings of SPIE

Soong TT (1988) State of the art review: active structural control in civil engineering. Eng Struct 10(2):74–84

Soong TT, Pitarresi JM (1987) Optimal design of active structures. Comput Appl Struct Eng 579–591

Teuffel P (2004) Entwerfen adaptiver strukturen. Doctoral dissertation, University of Stuttgart, ILEK, Struttgart

Tibert G (2002) Deployable tensegrity structures for space applications. Doctoral dissertation, Royal Institute of Technology, Stockholm

Utku S (1998) Theory of adaptive structures: incorporating intelligence into engineered products. CRC Press LLC, Boca Raban

Veuve NW, Safei SD, Smith IFC (2015) Deployment of a tensegrity footbridge. J Struct Eng 141(11):1–8

Vincent JFV (1990) Structural biomaterials. Princeton University Press, Princeton

Wada B, Fanson J, Crawley E (1990) Adaptive structures. J Intell Mater Syst Struct 1:157–174

Weilandt A (2007) Adaptivität bei Flächentragwerken. ILEK, University of Stuttgart, Stuttgart

Williams C (2000) British museum great court roof. http://people.bath.ac.uk/abscjkw/BritishMuse um/. Accessed 14 Apr 2013

Xu B, Wu S, Yokoyama K (2003) Neural networks for decentralized control of cable-stayed bridge. J Bridge Eng (ASCE) 8:229–236

Yao J (1972) Concept of structural control. ASCE J Struct Control 98:1567–1574

Ziegler F (2005) Computational aspects of structural shape control. Comput Struct 83:1191–1204

Zuk W, Clark RH (1970) Kinetic architecture. Van Nostrand Reinhold, New York

Chapter 9
A New Look at Robotics in Architecture: Embedding Behavior with Smart Materials

Doris K. Sung

Abstract Although the field of robot design goes back many years, the use and study of robotics in architecture is still relatively in its early stages. The reliance on hardwired or battery-powered electricity as well as the burden of installation and maintenance costs bring serious criticism to the use of these products. At the same time, similar kinetic systems that do not require electrical energy or Artificial Intelligence for instruction are being developed. These passive-active systems can provide redundancy and alternatives to the range of robotics available in buildings today.

9.1 Introduction

Typically, robotics is associated with the fields of electrical engineering and computer science. State-of-the-art batteries and complex electrical wiring energize the robots to operate while microchips store information ranging from operation guidelines to complex algorithms designed for increased learning behaviors and appropriate responses. However, there is a less common path of study that toys with similar kinetic systems, but do not require electrical energy or Artificial Intelligence for instruction. And when installed in architecture the inclusion of such systems can be both cost and operationally effective at completing critical tasks. To use terms familiar to architecture, these types of components can be considered "passive-active" in nature—"passive" because they use no electricity and "active" because they are physically responsive.

Although the field of robot design goes back many years in both the mimicking of the human form starting from its first appearance in Karel Capek's novel called R.U.R. (Capek 1923), the use and study of robotics in architecture is still relatively in

D. K. Sung (✉)
School of Architecture, University of Southern California, Los Angeles, USA
e-mail: doris@dosu-arch.com

D. K. Sung
DOSU Studio Architecture, Los Angeles, USA

© Springer International Publishing AG, part of Springer Nature 2018
H. Bier (ed.), *Robotic Building*, Springer Series in Adaptive Environments,
https://doi.org/10.1007/978-3-319-70866-9_9

its early stages. More popular in the fabrication process of architectural components, a generation of robotic devices are appearing in active façade designs fitted with Arduino actuators and other wired controls. The use of sensors in buildings is more and more commonplace where commercially available products such as the Nest thermostats, security systems, doorbells and other home automation devices are now standardized household gadgets. But the reliance on hardwired or battery-powered electricity as well as the burden of installation and maintenance costs bring serious criticism to the use of these products. What happens in a power outage? How can we afford and maintain wiring inside the building's walls? How can we identify replacement of microchips and train individuals on the maintenance of the chips? Given these simple concerns, passive-active systems can provide redundancy and alternatives to the range of robotics available in buildings today.

There are two basic ways to achieve passive-active response—using smart materials or designing mechanisms that directly respond to translatable natural forces. Because smart materials automatically respond to changes in their surroundings while using zero energy and no controls, their inclusion as an initial step in the design of dynamic systems makes reasonable sense. Available in many forms, these smart materials react reversibly to changes in the environment such as temperature, light, moisture, pH, stress, magnetic or electric fields. Ranging from shape memory alloys, piezoelectric materials, quantum-tunnelling composites, electroluminescent materials, to colour-change materials and more, these materials can be programmed to have very specific goals, which is very important when qualifying these systems under the category of robots. But programming here does not mean computer programming. In the analogue method, programming can be done materially in the manufacturing process, geometrically in the cutting process or formally in the fabricating process. And, if not using smart materials, static materials (both soft and hard) can be assembled to operate responsively as a translation of an environmental force such as wind.

Many might argue that the absence of computer controls and artificial intelligence in an automated mechanism is not a robot. But according to Maja J. Mataric's basic definition of robots in her book, *The Robotic Primer,* it clearly is a robot. She writes, "A robot is an autonomous system which exists in the physical world, can sense its environment, and can act on it to achieve some goal" (Mataric 2007) She goes on to break the definition down into five critical components that all robots must have: Autonomous System, Physical Embodiment, Environmental Sensor, Responsive Behavior and Specified Goals. The following describes these components in greater detail:

– Autonomous System: In order to be an autonomous system, the machine or device cannot be controlled by humans directly or by remote control. Some input from humans is acceptable, but complete control is not. For this specific reason remote-controlled vehicles do not qualify as robots. Roomba (a household robot vacuums), on the other hand, requires very little control so are considered robots. Responsive systems on buildings actuated by smart materials require no controls and are therefore automonous.

– Physical Embodiment: Having physical embodiment that exists in the physical world of unbendable physical laws and challenges is also a requirement if wanting to be called a robot. Many roboticists such as Mataric believe the robot must have a body. For them, simulated robots in cyberspace, or "bots", do not qualify. Adrienne Lafrance, a writer for the *Atlantic* magazine, clarifies: "Just as 'robot' was used as a metaphor to describe a vast array of automation in the material it's now often used to describe—wrongly, many roboticists told me—various automated tasks in computing. The web is crawling with robots programmed to perform tasks online, including chatbots, scraper bots, shopbots, and twitter bots. But those are bots, not robots. And there's a difference" (Lafrance 2016). For building parts, they are tangible and undeniably physical. Halograms, projections and other digitally-controlled parts of architecture exist and are not considered robots or part of this essay's passive-active systems.

– Environmental Sensor: The robot must sense or perceive its surroundings. It can be through touching, smelling, seeing, hearing, tasting, etc. but not through the delivery of information, as in the case of bots. Often in architectural applications, these sensors can be triggered by temperature changes, physical obstacles, rise in humidity, increase in wind speed, number of warm bodies or activity in the room. The popularity of using sensors in architecture has risen over the past few years in the development of gadgets for residential and commercial applications in both the buildings and in wearables that interface with the buildings around. Many of the sensors are digital, but in the case of smart materials, the material itself senses changes in the surrounding environment by changing some aspect of its physical properties. The indexing of this change is perceptible.

– Responsive Behavior: Taking action is a critical requirement for robotics. Once the robot receives sensory input, it must be able to respond by doing or changing something. The connection between input (sensing something) and output (acting on it) differentiates the robot from just being a sensor. It is in this area where the definition is ambiguous because action can mean many different things. But for smart materials the action can be part of the sensing mechanism or translated into some other response. Curling, twisting, changing color and shrinking are some of the possible responses.

– Specified Goals: Robots need specific goals to deem itself useful especially in a world where they are categorized in two basic areas: industrial and service robots. True to its original meaning, the robots are meant to do labor or hard work, but are differentiated by its relationship to people. According to the International Federation of Robotics (IFR), industrial robots are "automatically controlled, reprogrammable multipurpose manipulator programmable in three or more axes" (Weizmann Institute of Science 2017).[1] These robots are often found in areas of fabrication, assembly and industrial use. A service robot, on the other hand, "performs useful tasks for humans or equipment excluding industrial automation applications" (ISO 2017). These robots work closely with humans as do passive

---

[1] Wise Computing is an area of study where "the computer actually joins the development team as an equal partner—knowledgeable, concerned, and proactively responsible".

-active robots in architecture, so that when positioned in facades or collaborating in the assembly process like 'cobots' discussed later in this chapter, they are able to serve humans for a variety of purposes including cooling, shading, ventilating, strengthening, etc.

In H. James Wilson's report in the *Harvard Business Review* called "What Is a Robot, Anyway?", he follows the IFR definition of Industrial and Service Robots adding a potential growth in the area of ethics or cultures in what is called wise computing (Wilson 2015). But in the case of performance actuating devices in architecture, robots in structures or facades are difficult to categorize. For precision fabrication processes of construction and structuring, the robots can be called industrial robots. According to IFR's definition, robots that are not industrial robots are automatically considered service robots. But in the case of those identified in Peter Testa's *Robot House*, "where objects make objects" (Testa 2017), the inclusion of the robotic arm in architecture blurs the function (and sometimes the purpose) of the robot. On one hand, the robotic arm aids in the construction of the architecture while, on the other hand, it continuously morphs the shape of the space, alters the building's function or distorts the perception of the form. These changes can accommodate or dictate how humans perceive or occupy architecture and thereby could be consider a type of service robot. But if the robot arm is eliminated from the equation and replaced with an army of tiny simple robots populated across an entire building surface, one can argue that the multitude of environmentally responsive parts directly serve buildings. And because those same buildings serve the occupants, the matrix of minirobots by transitive property serve the occupants.

In architecture, the discussion of adaptive, responsive and interactive environments straddle an area between embedded computation and physical dynamism. It is the convergence of a digital and physical version of the idea of interaction. When the complexity of computation is programmed into the design of the pieces, the intelligence is embedded into the physical component. But rather than designing each piece to gain added intelligence like that in algorithms made for artificial intelligence, these simple architectural components are designed to be limited and focus only on its task at hand, sensitive to the slightest changes to its environment. It may be arguable whether the embedding process reveals or hides the computed intelligence. The discrete forms do not readily reveal the individual piece's programmed purpose because it oftentimes is masked by a larger tessellation pattern or systematic joining pattern. But upon seeing its calculated response to slight changes in the environment, the movement clearly indicates its intended yet limited purpose.

Highly performative and efficiently purposeful, the robotic projects on the following pages are designed to be simple, focusing on the bare essentials. And when meticulously engineered, the individual parts play a specific role to perform unique movements so that when strategically multiplied across a surface or form, they team up to produce a fluidre action. Each robot, performs its task tirelessly and endlessly. The obedient workers have no added intelligence, only embedded behavior. Although this chapter identifies three general task categories (self-shading, self-assembly and self-propulsion), it is by no means intended to limit the potential of smart materials (or programmable matter) in architecture or argue that it should replace wired robots (or artificial intelligence). Rather it is meant to complement the evergrowing body of robotic research in architecture and to offer a different approach to how we design and perceive architecture. These binary aspects cannot be developed in isolation and must synthetically inform each other as we look to a future of living with and amongst robots of all kinds whether they have a mind of their own or not.

9.2 Task One: Semi-soft Robotics for Self-shading

In response to the stiffness of standard articulated robots and robot arms, the emerging subfield of soft robotics uses soft and deformable materials and structures to perform dynamic tasks such as the grasp and manipulation of unknown objects, locomotion in rough terrains, and physical contact with living cells and human bodies. By using softer materials, the ability to manipulate a wider range of motion allows greater potential for control and versatility. Despite the challenges of establishing a new growing subfield, the possibilities for new materials, tools, uses, simulation technologies, testing methods and implementation are vast (IEEE RAS 2017). For Saul Griffith, cofounder and chief executive of Otherlab, designing various soft robots and systems are part of the future. "Every problem in mechanical engineering has been addressed with more weight, more power and more stiffness," said Mr. Griffith. "But nature—the real world—is squiggly" (Hardy 2015). His company uses pneumatics to make inflatable arms, robots, gas tanks and more. Other scientists are developing the use of softer materials in robotics with various success. Joey Davis Greer and his colleagues at Harvard University are investigating growth in soft robots. The creeping tentacles mimic the growth of vines and can elongate up to 1,000 times its original size at speeds up to 22 mph. Because the behavior of the softer material is not constricted by a limited number of configurations like hard articulated systems, the dynamic activity of these tentacles is continuously changing and its positions infinite (Farokhmanesh 2017). And, because the fluidity of positions incorporates movement or growth, the connection to nature is even more uncanny, which may explain designers' interest in looking to biology for inspiration or viewers' need to correlate the creeping motion to things found in nature.

For architecture and landscape architecture, the need for a connection to the environment may seem obvious, but the realization of such projects has only recently emerged due to the development of various software, materials and fabrication methodologies, as well as an interest to move away from sensor-laden gadgets. In the Foreward for the book, *Responsive Landscapes*, Jason Kelly Johnson and Nataly Gattegno point out the "conceptual shift from a more object-oriented understanding of technology as a mediator between systems to a more integrated and synthetic understanding of technology as the medium through which we can encode and amplify landscapes with intelligence and heuristic capacities" (Johnson and Gattegno 2016). This interest can be seen in various biophilic projects by architects where the constructs simulate natural landscapes, to ones that synthetically interface with the atmospheric components of nature such as air quality, temperature, humidity and wind. In the case of 'Reef', an installation at the Storefront for Art and Architecture in NYC, Rob Ley and Joshua Stein designed an interactive serpentine surface that offers people companionship like that of a robotic pet. The stereoscopic sensors triggera select number of the 600 + fins to curl by use of the smart material Nitinol, an alloy of nickel and titanium. When heated, the smart wire-like material shortens and distorts the cut sheet plastic into a curl. The resistance of that same plastic stretches the wire back into its original length as the temperature cools. Even though the surface looks like an array of fish scales, the overall effect when in motion is a fluid movement across the field of pieces much like flowing kelp in an ocean's current.

'Epiphyte Chamber' by Philip Beesley, another project that reacts to human movement, has a more immersive effect resembling that of the Pandora jungle in the movie Avatar. The imprecision of the soft materials, in this case, allows the unidirectional Arduino movement to misalign and deform in multiple directions. When hinged at various locations, the movement and gravity enhance the effect by swinging and bouncing. Combining both hard, articulated action with soft robotic movement and soft materials, the resulting installation creates diffusive boundaries and atmospheric qualities especially when presented in darkness and theatrical lighting. The more standard type of mechanical joining systems is combined with flitting of feather-like forms and curling of squid-like tentacles, all moving independently throughout the installation. Hidden sensors trigger responses that ripple through the installation in peristaltic waves (Beesley 2016). The combination of thousands of abiotic pieces uncannily references a buoyant and vibrant jungle bursting with lifeforms. Without the use of soft plastics and pliable rubber materials, it would be nearly impossible to mimic animal or plant movement in a manner like this installation.

Smart sheet materials, although not truly a soft material, can express the essence of biota in the physics of its movement behavior as a response to environmental changes and in its redundancy as a system. Because the movement is not mechanical and does rely on any articulated joints, smart sheet material for architecture can achieve very similar effects as soft robotics and can be called semi-soft. The motion is not controlled by computers but by the geometry the piece is cut. When cut in uniform widths, the resulting curvature is radial. But, when the cut is tapered, the curl is a parabolic or even more complex curve. Twisting and spiraling are also controllable movements with no additional technical equipment used for sensing, processing or actuating. This type of control of behavior can be seen in wood veneer projects such as 'Hygroscope' and 'Hygroskin' by Achim Menges (with Steffen Reichert and Oliver David Kreig). Thin sheets of quarter-cut veneer in combination with a synthetic fiber-reinforced polymer react to changes in humidity by incorporating a combination of active cell pressure and passive systems independent from a plant's metabolic trigger mechanism. As the moisture level in the atmosphere increases, "water molecules become bonded to or released by the material. The changing distance between the microfibrils within the wood cell tissue causes shrinkage and expansion" (Fox 2016). In both these projects, the triangular geometry of the pieces allows the fastest and greatest amount of curl towards the center of the hexagonal apertures, where airflow is less turbulent.

Fig. 9.1 The surface of this outdoor demonstrative pavilion is made of flaps that curl when heated allowing ventilation of hot air or shade from the sun

Similarly, smart thermobimetal curls in response to changes in its environment. Rather than responding to fluctuations in humidity, thermobimetal reacts to temperature changes. A molecular bonding lamination of two alloys of metal, each surface has different coefficients of expansion. As the temperature rises, one side expands more than the other resulting in a curling action. Because the material is capable of curling in multiple directions (the wood veneer curls uni-directionally), the exact geometry is critical in determining the final behavior of the piece and can be considered a method of programming behavior into this type of simplistic robot. Depending on the shape it is cut, each piece can be designed to curl at different tightnesses, to twist in designated directions or to resist curling altogether by imposing cross-directional internal forces. This specificity can provide surprising control when the material is heated above 80 °F by either ambient air temperature or by solar radiation (Fig. 9.1). Depending on the degree of angle to the sun, orientation on the site or function below the canopy, various parameters developed from numerous early behavioral test with the material can be inputted into the software during the design stage so that when the surface of the project is physically populated, individual parts can behave in a controlled manner, consolidating the response and action into a single movement.

In 'Bloom' the operation of the curl is in a cantilever configuration. Three sides of each crucifix-shaped 0.008" thick thermobimetal tile is connected to adjacent pieces in an interlocking weaving pattern while one arm is allowed to move freely.

In the design phase, this active arm is parametrically programmed to vary in length to react to the sun in a tight or loose curl depending on whether its function is to block the sunlight or release hot air from below. In areas where there is no need for functioning parts, the moving flap is decreased in size or made absent to save on material cost and overall structural weight. The other three arms of the crucifix are fixed at the exact length determined by the matrix with the exception of the flaps opposing arm. Allowed to vary in width, this upper arm controls the size of the opening for ventilation in a fixed position. Each of these parameters are determined by the connecting parametric software with digital analysis tools. The weave pattern of twenty-three thermobimetal tiles allow the 414 panels to take unique hypar (or hyperbolic paraboloid) forms and perform like localized structural shells. Because the structure is incorporated as the frame to each panel, the monocoque surface of undulating panels is easy to assembly without large structural members and without the need to use onsite heavy equipment. When completed, the complex double-curving surface is strong and lightweight.

In 'InVert' window shading systems, thousands of thin thermobimetal pieces react to solar radiation and simultaneously block the sun from entering the building (Fig. 9.2), but the pieces perform differently than those in the 'Bloom' project. When heated, each 0.0025" thick piece shifts its center of gravity to toggle from one position to another. For this purpose, the geometry of the pieces is designed to not only curl in a certain direction, their mass is calculated to throw its weight from one position to the next with a single fulcrum support and two cantilevering ends. This reaction, programmed by its geometry, is calibrated to operate at certain temperatures while at the same time the same geometry doubles to block the sunlight from entering the building at specific times of day and specific times of year. Optimization occurs on three levels: performance of the bimetal curl and flipping, temperature at which it responds (a combination of ambient temperature and solar absorption) and performance of shading at the hottest times of day during the hottest days of the year (Fig. 9.3).

Fig. 9.2 These timelapse sequences display the "InVert" self-shading system responding by blocking the sun from entering the building

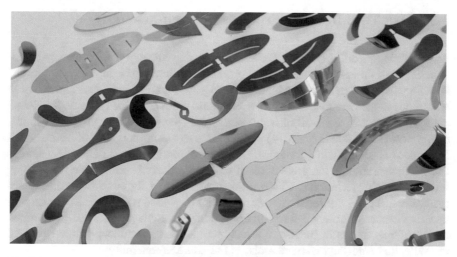

Fig. 9.3 One of the methods of programming behavior in thermobimetal is determined by the geometry of the piece so that they both flip at the correct temperature and shade at the right angle

In these special windows, the thermobimetal robotic pieces are positioned in the cavity of an insulated glass unit (IGU) to strategically allow sun to enter the building when the outdoor temperature is cool in the winter, but flips to block the sun when the temperature is hot and the sun is radiating in the summer. By automatically controlling the amount of solar heat gain from entering the building in the hot months of the year, the load on air conditioning is alleviated, reducing the energy usage and the hot air emission from a building into the atmosphere. Not only does it eliminate expensive wiring in the installation and maintenance of a building, it is a zero-energy and no control system that retains the same high level of visible transmittance and view when it is both open and closed. Additionally, low-e coatings or films are no longer needed so that the building's occupants can enjoy a greater range of color spectrum than ever before. The impact of daylight, view and color cannot be underestimated in the mental well-being of humans cooped up in indoor environments.

Because the windows are populated by a pixelated matrix of soft thermobimetal pieces where individual piece can operate independently of its adjacent neighbor, the combinations of flipped and not flipped pieces are multitudinous. Depending on the configuration at any time of day, the array will be indexing the temperature and climate of the outdoor environment. When a cloud moves across the sky, its movement will be tracked by the small pieces inside the IGU and the surface pattern will move across the surface of the building. Additionally, the matrix inside the windows can be configured as a gradient of wider and narrower pieces depending on the needs of the user for solar shading and the interest in increasing visible transmittance (daylight) and view. Like the projects by Ley/Stein and Beesley, these wave-like patterns bring life to the once static building skin and have the potential to connect humans emotionally to a building and to the surrounding atmospheric environment.

9.3   Task Two: Intuitive Processes for Self-assembly

As computing capabilities and material programming converge, interest in smart construction, ease-in-assembly and low-cost manufacturing is bringing out new innovations in the area of making and manufacturing. Robot arms and robotics have been ubiquitous in the manufacturing process of mass production for many decades with renewed interest in areas of mass customization (San Fratello et al. 2017), industrial robots (Gramazio et al. 2014) and the robotic accommodation of human error (Advanced Design Studies Program 2017) in the construction process. The continuous demonstration in the production of unique products is impressive when using rapid prototyping techniques, CNC milling machines and digital laser-cutting equipment. But there is also an area of study of wireless self-assembly processes that opens up new avenues of how we make things. Adding a new layer between the human-made and robotic-made, the use of smart materials can introduce a whole new type of fabrication aids (beyond hand tools) in the area self-assembly.

The path to self-assembly has degrees of autonomy. The first step towards intelligent assembly methods is designing products that do not require instructions or assembly information. With the use of the computer and clever identification markings, the form can emerge from the assembly of specific parts. Indoctrinated with early exposure to IKEA pictographic instructions and intuitive platforms in smartphones, designers are considering how fabrication methods in construction can eliminate instructions, training and specialized skills. When dealing with computer-driven tessellation projects where no two pieces are alike, the use of traditional architectural drawings are often obsolete. With most of the construction process simplified in the design process, the unique building components and panels are prepared by digital laser-cutting machines, CNC milling machines, 3-d printers and robot arms. Special numeration systems are incorporated to the design so that workers on site only need to match localized marked pieces together rather than assemble the project in a hierarchical manner. When the structure is incorporated in the project's panels, the need for large equipment eliminated and the joining system designed to be simple, the geometry of the pieces will ultimately determine the final shape of the installation. In the case of the 2011 and 2014 research pavilions at the ICD/ITKE in Stuttgart (ICD/ITKE 2011 and 2014), the panels have a specific position in the pavilion and when fully assembled, the final form can structurally perform properly.

In the project 'Bloom', extensive instructions commonly found in construction documents are completely eliminated. Instead, instructions are reduced to two simple mappings identifying the piece's or panel's location in the pavilion. Each panel is made up of 23 pieces of thermobimetal and four perimeter frames. With rows identified with alphabetical letters and columns by numbers, each piece is etched with a multi-digit code that would identify its position on the panel and the position of the panel on the overall pavilion. Only two drawings are utilized on site during construction. One is a drawing of the surface flattened and unrolled like a geographic mapping, while the other is a 3-d printed model of the pavilion with the multi-digit code visible on each panel (Fig. 9.4). With these two "drawings", the use of the computer is also unnecessary at the site. The entire pavilion is assembled with rivets and nut/bolt construction. No skilled labor is required.

The completion of the 'Bloom' pavilion raises the question of thermobimetal's potential role in the assembly process as an active construction partner. In the current system, the relationship between humans and tools is direct when constructing architecture or assembling parts. There are no other means of construction aid involved in the process unless smart materials are introduced. If incorporated strategically, materials like thermobimetal can aid in the process and make assembly easier. They can both perform as micro industrial robots during the assembly sequence *and* as the building material in its final resting position.

For the design of a pre-tensioned lightweight surface called 'eXo' (Fig. 9.5), the thermobimetal pieces are heated to a higher temperature and accurately curled to slide into its designated position without force during the assembly process. In this project, which is the bottom of a five-tiered 40″ tall tower, the 0.03″ thick thermobimetal pieces, too difficult to mechanically curl by hand but stiff enough to support higher loads, are heated in an industrial oven to about 250 °F. The long pieces, operating as a beam and supported at both ends, are then quickly and easily inserted into position using a single gloved hand. No added physical force was needed. As the piece cools, it starts to flatten, pulling the aluminum straps into tension (Fig. 9.6). The straps are designed to prevent the thermobimetal from completely flattening, constricting its final configuration into the shape of an archer's bow and capturing the latent forces into potential energy. When multiplied into a matrix in the form of a surface, the lightweight mat is super strong and can be applied at many different technical scales from nano to macro. Not only does this method eliminate the use of hand tools and brute force, the thermobimetal performs as a robotic assistant in this process before arriving at its final bowed position. It cannot operate without human intervention, but its curling behavior is critical to the functioning of this relationship. And, if the human in this case is replaced by the robot arm, the smart material will still retain its job as a micro industrial helper.

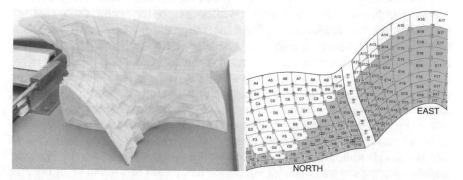

Fig. 9.4 Because the assembly method of 'Bloom' was incorporated in the design phase, only two construction documents were used on site: a mapping of the panels and a 3D print of the final form

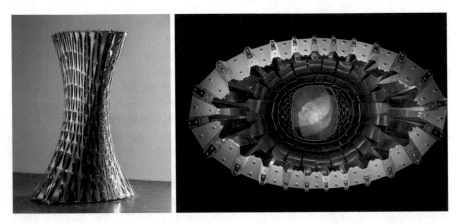

Fig. 9.5 This pre-tensioned structural tube called 'eXo' is the bottom tier of a five-tiered tower

Fig. 9.6 The sequencing of assembly of the 'eXo' project combines thermobimetal springs with aluminum straps

An increase in the role of the active abiotic subcontractors like thermobimetal leads to the possibility of assembly methods that needs no hands and no humans. In the development of self-assembly in architecture, smart materials, programmable materials, geometry, and computation play large roles. According to George M. Whitesides and Bartosz Brzybowki in their article "Self Assembly at All Scales", they identified self-assembly as "the autonomous organization of components into patterns or structures without human intervention" (Whitesides and Brzybowki 2002). This definition is further clarified by Skylar Tibbets as "the process by which disordered parts build an ordered structure through only local interaction" (Tibbets 2017). Instructing or programming robots to build themselves seems like an obvious next step to the development of computer-programmed robots (Von Neumann 1966). However, when working without electricity (or batteries) and no computer controls (or algorithms), the idea of self-assembly requires resourcefulness in how this is actually done.

Although the incorporation of self-assembly into general architectural applications may seem distant, the explorations in this area of study expose new ways of thinking. Eliminating wiring by finding other airborne energy sources and using smart materials are ways to make this happen. A team of researchers at the Wyss Institute for Biologically Inspired Engineering and the John A. Paulson School of Engineering and Applied Sciences (SEAS) at Harvard University has created battery-free folding robots. They are able to "demonstrate a battery-free wireless folding method for dynamic multijoint structures, achieving addressable folding motions—both individual and collective folding—using only basic passive electronic components on the device. The method is based on electromagnetic power transmission and resonance selectivity for actuation of resistive shape memory alloy actuators (Nitinol) without the need for physical connection or line of sight." (Boyvat 2017) Nitinol is a smart wire material that shrinks when heated. When doubled with some physical resistance, it can automatically assume its original shape. Using simple origami patterns and smart Nitinol, the Wyss team has devised small hand-sized robots to grip and contort. For microrobots, this is significant because it eliminates the reliance on bulky batteries that hinder the development, movement and operation of small-scale robots.

In the Self-Assembly Lab at MIT, Skylar Tibbets works with students from various backgrounds seeking "robotics-like behavior without the reliance on complex electro-mechanical devices." (Tibbets 2017). He deposits materials such as hydrogel alongside other rigid materials so that when submerged in water, the hydrogel swells causing the composite to curl or fold. The geometry of the composite can be controlled by adding disks or other elements at the point of movement to precisely transform into a variety of shapes. The junctures can be combined to produce other reaction like expansion or contraction. The projects range from self-assembling geodesic spheres to warping landscape-like surfaces. And, with the Programmable Matter[2] team, the Self-Assembly Lab has developed self-transforming carbon fiber, printed wood, custom textile composites and other rubbers/plastics. In the case of the programmable printed wood, a wood composite filament sensitive to changes in humidity is 3D printed into a bilayer where one side is smooth and the other side corrugated for easy bending. When exposed to moisture, the smart material can fold and curl by itself into a variety of three-dimensional shapes. And when dried it returns to its original form.

For materials that respond to changes in temperature like thermobimetal, the curling response can be systematically manipulated to performs tasks like link a chain together or transform a flat-packed system into a volumetric shape (Fig. 9.7 Link). Again, here the two-dimensional geometry plays a critical role in how the form assembles itself. Like previously described projects, this material behaves as both an assembly assistant and a building material simultaneously. It remains dormant until the temperature rises above 80 °F. As the temperature continues to rise, the linear strip curls back on itself to close the link much like a clasp on a necklace. The pieces can work sequentially or simultaneously, in a line or in a matrix, and at multiple scales. Potential use for such projects can be for inflatable textiles or architectural surfaces for insulation at both microscales and macroscales. Work in this area continues to develop for future application.

When applied to larger scale elements, the assembly process can be designed to be one way and incorporate a locking mechanism to be used in places that are hard to reach or difficult to travel to such as remote locations as far as the moon or underwater.

---

[2]Programmable Matter, according to Tibbets, is the study of physical matter that has the ability to change precise form and/or function by design.

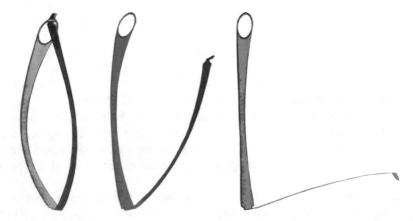

Fig. 9.7 This self-clasping link of thermobimetal is only 2″ long

Using only common nuts/bolts as resourceful weights to freeze the assembly process into its final form, the originally flat crucifix shape (Fig. 9.8 Box) can erect itself to a four-inch height when heated to the right temperature. On a systematic level, the dense x-y matrix of these cross-shaped pieces can transform a scaled-down suburban landscape to the z-direction as mini-towers in the park. This type of system can be scaled down to a micro level to make surfaces patterns translate from translucent to opaque and vice versa. Upon the removal of the gravity-based weight, a flattened pentagonal thermobimetal system can swell into a spherical wheel shape (Fig. 9.9 Sphere) and stay in place by an internally oppositional thermobimetal mechanism. As the temperature rises and the wheel self-inflates, the internal framework locks the volume into place and prevents it from returning to its flattened form. Unlike the previous version, this one-way form does not depend on gravity to lock it into place, opening up the possibility of using this type of system in atmospheres that are more buoyant than that of Earth. This self-inflating wheel is currently being incorporated on a smart toy vehicle by the author. When set in the hot sun long enough, an entire thermobimetal vehicle will be able to erect itself and propel forward by a separatesolar-powered mechanism. Hopefully, this flat-packing and self-assembly process will also contribute to the alleviation of problems in packing/shipping industry and in the reduction of carbon footprint.

9.4 Task Three: Personification of Self-propulsion

As tectonics begin to incorporate more and more robotic elements that move in quirky ways, humans instinctively associate the movements to personality traits which then lead to stronger personal bonds. Depending on how they are designed purposefully or accidentally, their behavior can express anger, playfulness, awkwardness, surprise, laziness, etc. They can limp, shuffle, drag their feet and be constantly tripping. Their movement can resemble that of biological plants, animals or other organic lifeform. Identifying something familiar in the unfamiliar is a natural response for humans. But going beyond and making an emotional connection adds an additional level of responsibility to the designer especially in the development of service robots.

Several decades ago when computer controls were starting to be integrated into machinery for automated functions (e.g. washing machines, automatic doors, etc.), the natural human response was to consider them 'robots'. Anything automated and foreign was called a robot. "When new technologies get introduced, because they're unfamiliar to us, we look for metaphors. Maybe it's easy to draw metaphors to robots because we have a conceptive model in our mind…I don't know if it's that they stop being robots; it's that once when we find comfort in the technology, we don't need the metaphor anymore" (Lafrance 2016). As these 'robots' were replaced with other machines that were modeled after the human body, they lost the distinction of being called robots. Only machines in the shape of humans or parts that looked like human parts (the articulated robot arm) were subsequently called robots.

These early robots were difficult for humans to trust and bond with. Many found them creepy and frightening. Their uncanny resemblance to humans as described by the psychologist, Sigmund Freud, in *Das Unheimliche* (The Uncanny) was psychologically disturbing and difficult for humans to connect with. Like automatons,[3] they were ghosts in machines. This reputation was further marred by the movie industry. Even though the robot's machine characteristics were erased in movies, it was replaced by enough evilness to engender even more mistrust. Cyborgs and androids in science fiction movies(e.g. Arnold Schwarzenegger in *Terminator* or the character Bishop in *Aliens*) looked and acted more human than ever before. But they often depicted villains and their likeability factor was very low. To offset this association and make the public feel comfortable with service robots for everyday use, scientists needed to design the robot to be likeable for humans to connect and to

Fig. 9.8 This self-erecting box reaches over 4″ tall when complete

---

[3]Definition of Automaton: a mechanism that is relatively self-operating; a machine or control mechanism designed to follow automatically a predetermined sequence of operations or respond to encoded instructions; or, an individual who acts in a mechanical fashion (*Merriam-Webster Dictionary* 2017).

Fig. 9.9 This self-erecting sphere can be made into a wheel for a toy vehicle

emphatize with the machines. In the case of Google self-driving cars, an adorable, cute car is easier to like than a powerful human-shaped robot. "The infantilization of technology is a way of reinforcing social hierarchy: Humankind is clearly in charge, with sweet-looking technologies obviously beneath them" (Lafrance 2016). Some strategies purposefully make robots look dumb so that they are likeable and not evil. Take Baymax, the inflatable robot in *Big Hero 6*, for example. He is squishable, huggable and loveable. These aspects make him easily approachable where users can feel emotionally connected to him, which is another reason soft robotics are becoming more popular.

Google's patent (US8, 996, 429) for a robot, whose personality can change depending on the user's mood, gathers data by correlating weather data to previous moods or connect a person's diction or word pattern in emails to estimate that user's mental status. The idea of likeability is taken one step further so that this type of robot becomes an extension of one's own personality. In many ways, the artificial intelligence in the movie *Her* took the notion of likeability to an extreme. The main character, Theodore, fell deeply in love with the personal assistant interface in his computer's operating system that shows similarities to Apple's "Siri". This cautionary tale expresses the power that intangible bots and possibly physical robots can have on humans especially when emotions are involved. But it is important for humans to think of robots as helpers and not companions. "As autonomous systems become more sophisticated, the connection between input (the programmer's command) and output (how the robot behaves) will become increasingly opaque to people, and may eventually be misinterpreted as free will" (Lafrance 2016). Everyone must keep in mind what is the function, not the form of a robot. Otherwise, the separation of biota and a biota become even more blurred.

For thermobimetal Critters, the design intent is not to create service robots or even robots with personalities. Early designs are intended to simply test the use of thermobimetal for the purpose of propulsion using the sun's heat to activate the movement. The design of these critters are valuable exercises on testing cantilever arms, gravity manipulations, structural folds, trim cuts for increase friction, and complex movements. Completely disconnected from the scientific process, the design of these hand-held robots are driven by unrestricted intuition and years of experience with the smart material. When the need to finely calculate and meticulously document the steps of design are removed, the liberating design process can produce some very creative work. As the collection of critters grow, the personalities of the mini-robots produces surface and the evaluation of the individual critters are based on likeability or cuteness because even the designers are human. Over time, it is impossible not to treat them like pets and even refer to them with nicknames. So, naturally, Rolie-polie somersaults, Silverback skootches, Mr. Froggie hops, Duckie waddles and Dorothy Hamill spins. To increase comraderie between the office's employees and their mobile robots, once-a-week races albeit ultra slow (sometimes, the critters only take one step on clear days) encourages loyalty not dissimilar to spectator sports fans and their teams. Clearly, the infantilization of technology of these self-propelling robots produces very strong hierarchical bonds between maker and product or between human and robot. Eventually, the knowledge gained from these exercises will inform the further development of the self-assembled and self-propelled vehicle described earlier, especially after the scientific process is applied.

The same bonds exist in the low-tech walking structures designed by Theo Jansen, an engineer turned artist in the Netherlands. Since 1990 he has been building mechanisms the size of elephants called 'Strandbeest' that walk when powered by wind. Predominantly built from yellow plastic polyvinyl chloride (PVC) tubes, these self-propelling beasts resemble a variety of animals with distinct personalities—lumbering gaits, writhing bodies and all. As with the thermobimetal projects, precise calculations and meticulous care to details are invested in the design phase so that the programmed articulation at each joint contributes to the overall walking mannerisms. The redundancy in the number of legs may seem extensive, but each one is essential and engineered to support smooth operation. Strength of wind, durability of the plastic, human error in the fabrication of the joints and unstableness of the earth give the beastsless-than-perfect movements, similar to humans. The imperfections give them character and make them lovable. Even Mr. Jansen refers to the mechanical animals and their habitat as if they are alive: "The sandpit is the pre-heaven for the beach animals. They are not yet ready to survive the real beach. I still have to train them. Usually I take them out once a year to the real beach to let them get a taste of their natural environment" (Theo Jansen's Strandbeest 2017). Alive or not, these bonds tie.

Unlike the previous two tasks, it is currently difficult to project how self-propelling robots can influence architecture on a tectonic level, but the personality that comes out of robotic movement will inevitably affect the way we view architecture and possibly even change our relationships to a building on an emotional level. When parts on a building are moving on their own without resembling articulated mechanical movement seen in automatons or industrial robotic arms, will we become more emotionally attached to the buildings? Will we care for the building like a pet? Can we love, hate, be saddened, be frightened or be surprised by the building? This new question of personality will become a new category of criteria that architects and engineers can no longer ignore and must strategically incorporate in the process of designing buildings with responsive parts.

9.5   Conclusion

The projects in this chapter are selected to illustrate the impact of passive-active systems on robotics and on architecture. The capabilities of the elements may have limited intelligence, but when considering society's interest in seeking zero-energy solutions, the use of zero-energy materials seems like a reasonable alternative and complement to the growing number of active systems in architecture with artificial intelligence. In a society obsessed and sometimes burdened with the ubiquitous presence of the computer, it is a refreshing opportunity to reflect on our anthropogenic impact on this world. With so many choices of smart materials and programmable matter available (and more being developed), the way they will change how we design architecture and other products asks more questions than provide answers. They make us reconsider what architecture is and imagine its potential roles in making our changing environment and evolving culture more robust.

But to design smart materials and other passive-active systems to behave and perform their special tasks, it is important to highlight the method of development. "Post-digital designers more often design by manipulation than be determinism, and what is designed has become more curious, intuitive, speculative and experimental" (Shiel 2008). Although top-down methods of design are commonplace and oftentimes necessary in the design of large scale urban projects, it is the bottom-up method that liberates creativity in these relatively small scale projects. At the beginning stages of each thermobimetal project described in this chapter, early tests with various geometries (cutting methods, folding techniques, joining details, heating sources) when done in the strictest of scientific methods result in the development of new methods of design that are repeatable. These early steps are necessary when dealing with new materials for both the internal development of the design inside a single design office and the greater design and scientific industry as a whole. "Designing has become a liquid discipline pouring into domains that for centuries have been the sole possession of others, such as mathematicians, neurologists, geneticists, artists and manufacturers" (Shiel 2008). The blending of the design (Cross 2011) and scientific methods are becoming less distinguishable. And, when there is no final goal or application in sight, the early studies remain idle in a vast library of information until it is resurrected for another small step forward. Little by little, the projects have emerged from an initial performance criteria to a simple geometry to a method of array to a tessellation pattern to a form and so on. At each level with each change, the pieces are put through rigorous tests so that they can be reproducible. "The complexity embedded in the design of responsive technologies requires iterative prototyping and computational development. This process of prototyping requires rigorous methods of making to tune sensing, feedback and actuation" (Cantrell and Justine Holzman 2016). Many of the projects shown are at different levels of completion. Some are at a very basic science level while others have been developed for specific architectural application. Design is applied at every time-consuming step of the scientific process. It is no surprise the projects are never completely finished.

It is at this point that we can project the impact of these materials to larger ecologies like the urban environment or the larger landscapes. Clearly, this anthropogenic process of designing inorganic moving parts on buildings as if they are alive is an anomaly in itself. What role does it play in the blurry area between biotic and abiotic systems? How will it shape our environments and our world? What will be the psychological impact on our society? Will we be able to differentiate between living and non-living beings? Brian Cantrell and Justine Holzman in their book *Responsive Landscapes: Strategies for Responsive Technologies in Landscape Architecture* suggest that the combination can possibly be a blend of the three: "The landscapes that we can begin to imagine have the capacity to not only embed themselves within their context but can also evolve with a life of their own, a synthesis between the biological, mechanical, and computational" (Cantrell and Justine Holzman 2016). This unknown future opens up new ways for us to rethink our urban terrains where the design of responsiveness of materials contributes to "a rubric…to horizontalize the relations between humans, biota and abiota" (Bennett 2010). In the past, this equal positioning was considered an impediment to the ecological growth of an area. But, when considering the connection between the man-made interior of a building and the outdoor climate as ecology, these responsive materials can contribute in more positive ways. "In other words, when landscapes get hybridized with responsive technologies, they will have the capacity to better process and respond to the variable and multi-scalar inputs from their environments" (Johnson and Gattegno 2016).

But perhaps it is exactly in this area of synthesis where our desire to grasp and control the mystery of nature becomes enthralling and new ideas can be revealed. In a way, the same connections between science and art, between nature and man-made, between the poetic and fact and between something that is self-reproducing (autopoiesis) such as artificial intelligence and something that depends on others (allopoiesis) such as the projects in this essay resemble the binary distinction of Felix Guattari and GillesDeleuze's "machines desirantes" (Guattari and Deleuze 1972). This unconscious need for delight drives humans to continuously and endlessly connect and synthesize various relationships with one another and with objects wrought with emotion. Take the cabinets of curiosities in the 17th and 18th centuries and the Wunderkammern of the Renaissance, for example. "Those immense collections of "rare" objects, where the natural and the artificial—products of "divine" and human craft, respectively—lived side-by-side as objects of amazement." (Olalquiaga 2005/2006) The juxtapositioning of vastly different items of wonder exposed new ways of thinking and an insatiating curiosity (or desire) to understand the mysteries and puzzles of our world. "The cabinet of curiosities offers a parallel to the interlocking dynamics of the contemporary universe. Because it tightly encases a variety of wonder, it flattens hierarchies and allows new attachments to spring up. Plucking things out of the customary family contexts and inserting them into a space of invention does not represent a "clean break" with an organic method of filiation. On the contrary, it extends to each of us the creative opportunity for inventing further relations of our own" (Stafford and Terpak 2002). In the case of smart materials, the amazement lies in the ability for the materials to respond to stimuli without human intervention on a phenomenological level. And, despite the scientific explanation of the actuation, the reaction of humans is one of earnest wonder, which Stephen Greenblatt in his book *Marvelous Possessions: The Wonder of the New World* says "stands for all that cannot be understood, that can scarcely be believed. It calls attention to the problem of credibility and at the same time insists upon the undeniability, the exigency of experience." (Greenblatt 1992). Once again, we arrive in an empowering position to understand our world but not to conquer or control. Rather, it is important for us to seek synthesis between the dialectical worlds we live in—robots or not.

Acknowledgements This chapter has profited from the contribution of several team members and collaborators working on the presented projects:
BLOOM: Doris Sung (PI), Dylan Wood, Kristi Butterworth, Ali Chen, Renata Ganis, Derek Greene, Julia Michalski, Sayo Morinaga, Evan Shieh, Garrett Helm, Derek Greene, Kelly Wong (design team), IngalillWahlroos-Ritter (Design Consultant), Matthew Melnyk (Structure Consultant)
INVERT: Doris Sung (PI), Adelfrid Ramirez, Stephanie Truong, Justin Kang, Ramsey Young, Belinda Pak, Thomas Gin (design team), Karen Sabath, Scott Horwitz (commercialization team), Joon-Ho Choi (USC), Steve Selkowitz (Lawrence Berkeley Labs), Thomas Auer (Transolar) (engineering consultants)
EXO: Doris Sung (PI), Dylan Wood, Hannah Woo, Evan Shieh, Jessica Chang, Dennis Chow, Carter Shaw (design team), RoelSchierbeek, Gregory Nielsen, Laura Min (ARUP) (structural engineering), Alex Rasmussen (Neal Feay Co.) (fabricator)
LINK: Doris Sung (PI), Justin Kang (design team)
BOX: Doris Sung (PI), Isaac Chen (design team)
SPHERE: Doris Sung (PI), Thomas Gin, Ryan Thomas, Geoffrey Ford (design team)
CRITTERS: Doris Sung (PI), Elizabeth Phillips(design team)
Photo Credits: Derek Greene, Brandon Shigeta (Fig. 9.1), DOSU Studio Architecture (Figs. 9.2, 9.3, 9.4, 9.6, 9.7, 9.8, 9.9) and Alex Blair (Fig. 9.5).

References

Advanced Design Studies Program headed by Yusuke Obuchi and Kengo Kuma, Department of Architecture, University of Tokyo (2017), http://arch.t.u-tokyo.ac.jp/activity/ninety-nine-failures/. Accessed 03 Oct 2017

Beesley P (2016) The epiphyte chamber: responsive architecture and dissipative design. In: The Routledge companion to biology in art and architecture. Routledge, London, p 194

Bennett J (2010) Vitrant matter: a political ecology of things. Duke University Press, London, p 9

Boyvat M, Koh J-S, Wood RJ (2017) Addressable wireless actuation for multijoint folding robots and devices. Sci Robot 2(8) eaan1544

Cantrell B, Holzman J (2016) A paradigm shift. responsive landscapes: strategies for responsive technologies in landscape architecture. Routledge, New York, p 15

Capek K (1923) R.U.R. or Rosum's universal robots. Oxford University Press, London

Cross N (2011) Design thinking: understanding how designers think and work. Bloomsbury, London

Farokhmanesh M (2017) These soft robots are inspired by plants and move like sentient vines. The Verge, July 19, https://www.theverge.com/2017/7/19/15983174/soft-robot-plant-science-robotics-greer. Accessed 15 Oct 2017

Fortmeyer R, Linn CD. Kinetic architecture: designs for active envelopes. Images Publishing Group, Mulgrave, Victoria

Fox M (ed) (2016) Interactive architecture: adaptive world. Princeton Architectural Press, New York, p 125

Fratello S, Virginia GL, Fornes M presenters (2017) International symposium of mass customization and design democratization. Philadelphia, PA, May 12–13, http://sites.psu.edu/mcdd/. Accessed 03 Oct 2017

Gramazio F, Kohler M, Willmann J (2014) The robotic touch: how robots change architecture. Park Books, Zurich

Greenblatt S (1992) Marvelous possessions: the wonder of the new world. University of Chicago Press, Chicago

Guattari F, Deleuze G (1972) L'anti-Oedipe: Capitalisme et Schizophrenie. Les Editions de Minuit, Paris

Hardy Q (2015) The robotics inventors who are trying to take the 'hard' out of hardware. The New York Times, April 14, https://www.nytimes.com/2015/04/15/technology/the-robotics-inventors-who-are-trying-to-take-the-hard-out-of-hardware.html?mcubz=1. Accessed 15 Oct 2017

ICD/ITKE at University of Stuttgart.2011 and 2014 Research Projects, http://icd.uni-stuttgart.de/?p=6553 and http://icd.uni-stuttgart.de/?p=11187. Accessed 03 Oct 2017

IEEE RAS Institute of Electrical and Electronics Engineers Robotics and Animation Society (RAS), Technical Committee for Soft Robotics. 2017, http://www.ieee-ras.org/soft-robotics. Accessed 25 Sept 2017

ISO 8373:2012(en) Robots and Robotic devices vocabulary, https://www.iso.org/obp/ui/#iso:std:iso:8373:ed-2:v1:en. Accessed 17 Oct 2017

Johnson JK, Nataly G (2016) Foreword: towards a robotic ecology. In: Bradley C, Justine H (eds) Responsive landscapes: strategies for responsive technologies in landscape architecture. Routledge, London, pp xvii–xix

Lafrance, Adrienne. 2016. What is a Robot. *The Atlantic*, March 22

Mataric MJ (2007) The robotic primer. MIT Press, Cambridge, p 2

Merriam-Webster Dictionary. Definition of Automaton, https://www.merriam-webster.com/dictionary/automaton. Accessed 03 Oct 2017

Olalquiaga, Celeste (2005/6) Object Lesson/ Transitional Object. Cabinet Magazine, Issue 20 Ruins, Winter, http://www.cabinetmagazine.org/issues/20/olalquiaga.php. Accessed 15 Oct 2017

Shiel B (2008) Introduction. protoarchitecture: between the analogue and the digital, architectural design. Wiley, London, p 7

Stafford BM, Terpak F (2002) Devices of wonder: from the world in a box to images on the screen. The Getty Research Institute Publications Program, Los Angeles, pp 5–6

Testa P (2017) Robot house. Thames & Hudson, New York, p 27

Theo Jansen's Strandbeest (2017), http://www.strandbeest.com/beests_wind.ph. Accessed 01 Oct 2017

Tibbets S (2017) Self-assembly lab: experiments in programming matter. Routledge, New York, p 9

Von Neumann J (1966) Theory of self-reproducing automata. University of Illinois Press, Urbana and London

Weizmann Institute of Science homepage, http://www.weizmann.ac.il/pages/wise-computing. Accessed 25 Sept 2017

Whitesides GM, Bartosz B (2002) Self-assembly at all scales. Science. 295(5564): 2418–2421

Wilson HJ (2015) What is a robot, anyway? Harvard business review

Printed in the United States
By Bookmasters